MAKING A GRADE

Making a Grade

Victorian Examinations and the Rise of Standardized Testing

JAMES ELWICK

UNIVERSITY OF TORONTO PRESS
Toronto Buffalo London

Toronto Buffalo London
utorontopress.com

ISBN 978-1-4875-0893-7 (cloth)
ISBN 978-1-4875-3935-1 (EPUB)
ISBN 978-1-4875-3934-4 (PDF)

Library and Archives Canada Cataloguing in Publication

Title: Making a grade : Victorian examinations and the rise of standardized testing / James Elwick.
Names: Elwick, James, 1973– author.
Description: Includes bibliographical references and index.
Identifiers: Canadiana (print) 20210118261 | Canadiana (ebook) 20210118326 | ISBN 9781487508937 (cloth) | ISBN 9781487539351 (EPUB) | ISBN 9781487539344 (PDF)
Subjects: LCSH: Educational tests and measurements – Great Britain – History – 19th century. | LCSH: Examinations – Great Britain – History – 19th century. | LCSH: Education – Standards – Great Britain – History – 19th century.
Classification: LCC LB3056.G7 E49 2021 | DDC 371.26/2094109034 – dc23

This book has been published with the help of a grant from the Federation for the Humanities and Social Sciences, through the Awards to Scholarly Publications Program, using funds provided by the Social Sciences and Humanities Research Council of Canada.

University of Toronto Press acknowledges the financial assistance to its publishing program of the Canada Council for the Arts and the Ontario Arts Council, an agency of the Government of Ontario.

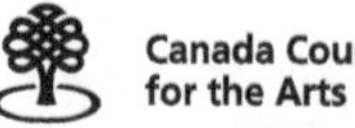

Funded by the Government of Canada | Financé par le gouvernement du Canada | Canada

For Sue Heddle, my own success coach

Contents

Figures

Preface and Acknowledgments

The saying that some books are not finished but abandoned is appropriate for this project. Some abandonment, some book: it's been a long process, made all the longer because of the numerous pathways wandered down then doubled back on. This wasn't because those routes were fruitless: quite the opposite. I hope other scholars can follow them. Other immensely suggestive realms include the link between exams and credentialism; links between French examinations such as the *concours* and British ones; examinations and the "Cardiff School's" studies of expertise and experience; more focus on non-English parts of the British Isles, particularly Scotland; the dynamics of rules and codification as well as the sociology of formalization; and the passage of forms of examination between China, the rest of Asia, Europe, and the British Empire.

Over these years I have amassed great debts. My greatest is to Bernard Lightman, for his mentorship and knowledge of other scholars in the field. One such colleague is Efram Sera-Shriar, whose Babbage-like organizational wizardry I imitated in delegating tasks (particularly to my transcribers, David Zatzman, Julia Nguyen, and Bernhard Isopp). Their work made it possible to range across a wider range of material. Another scholar is Ruth Barton, whose impeccable work is a model of precision I have sought to live up to.

Colleagues who read the entire manuscript were Ke Zunke, Malte Ziewitz, Hsiang-Fu Huang, Elihu Gerson, and Tina Choi. I am also indebted to the three referees, especially for critical comments on points I strove to correct. The University of Toronto Press's Stephen Jones's enthusiasm for this manuscript revived my own passion for it. Ellie Barton's copy edit produced greater clarity. Janet Friskney's skillful coaching at crafting applications helped me receive a Social Science and Humanities Council of Canada Insight Development grant.

I also received support from York University's internal research grants. I thank them all.

I am grateful for archival help from Karen Morgan at the North London Collegiate School; Jane Harrison at the Royal Institution, London; Anne Barrett at the College Archives, Imperial College; Gillian Cooke at Cambridge Assessment; and Tansy Barton, Angela Craft, and Ruth Macleod at the University of London Library. Geoff Belknap and Julie Hipperson obtained important pictures for me when I couldn't be at the archive in person. My work on the John Tyndall Correspondence Project was also very important in gaining a more precise sense of many of the figures discussed herein. I have worked with over a hundred members since 2006, and wish to thank them, particularly acknowledging the contributions of Daniel Zizzamia, Reut Ullman, Kylie Stasila, Tatyana Shestakov, Laurie Schell, Nedra Rodrigo, Sheri Repucci, Michael Reidy, Evan Radford, Sylvia Nickerson, Elizabeth Neswald, Zemina Meghji, Michael Laurentius, Nanna Kaalund, Frank James, Roland Jackson, Nicolas Sanchez Guerrero, Erin Grosjean, Julie Goriounov, Rachel Mason Dentinger, Gowan Dawson, Dave Cross, Ken Corbett, Rosanna Chowdhury, Geoffrey Cantor, Steve Bunn, Janet Browne, William Brock, Michael Barton, Micah Anshan, Michael Anderson, and Mark Ambrogio.

Other colleagues to thank for comments, suggestions, and stimulating conversation include Caitlyn Wylie, Paul White, Andrew Warwick, Joanna Szurmak, Matthew Stanley, Catherine Sloan, Anne Shteir, Simon Schaffer, Marsha Richmond, Chris Renwick, Theodore Porter, Allan Olley, Lynn Nyhart, Sylvia Nickerson, Hélène Mialet, Robin Metcalfe, Ben Marsden, Frederik Jonsson, Richard Jarrell, Martha Jarrell, Hillary Hutchinson, Hal Hansen, Ian Hacking, Graeme Gooday, Aileen Fyfe, Maurizio Esposito, Anna de Salvo, Melinda Baldwin, and Jon Agar. I wish I had spoken with the late William Clark. Part of this book draws on my chapter "Economies of Scales: Evolutionary Naturalists and the Victorian Examination Mania," in *Victorian Scientific Naturalism: Community, Identity, Continuity*, edited by Gowan Dawson and Bernard Lightman (University of Chicago Press, 2014), and this larger project has been polished and sharpened in discussions and seminars that gave rise to that paper. I thank all participants. Above all, however, I am grateful to Sue, Sydney, and Spencer for their support: miigwetch.

MAKING A GRADE

Introduction

"Standardized testing." Although we pay more attention to the second word, it's the standardization that makes such testing powerful. Standardization makes it possible for candidates at different locations to answer identical questions. Standardization makes possible the metrics and statistics that give public servants the information – or a belief they have it – to make decisions affecting others' lives. Standardization makes such tests into seemingly valid comparisons of knowledge, skill, and competence. And standardization is necessary for any examination's "grade" to be made level and seen as fair – a shared belief necessary for any meritocracy, if it is to persist. Thus where others have asked whether standardized tests are effective, or just, at assessing knowledge and skill, this book asks different questions that focus on the tests' uniformity. What gives people the reason to believe they're being assessed under the same conditions, by the same exam? What allows us – and our cultures at large – to trust the results of such tests? And how did some topics, and not others, come to be seen as more important and hence examinable?

When people take a standardized test in our own time, there are usually certain objects that are employed: uniform test booklets, Scantron® sheets, or digital screens. And there are techniques, rules, and routines – for instance, each candidate must sit a specified distance apart from another, be undisturbed, and have their answers graded, usually with numbers. These techniques, rules, and routines are called *repertoires*[1] – collective interactions among people that come to be learned and shared. This word is more precise than "culture." Collectively, these material objects and repertoires of testing are called *infrastructure* in this book. Its argument is that the infrastructure of examinations makes possible not only standardized testing, but also trust in the results. This book therefore uses the tactic of what the information management scholar

Geoffrey Bowker calls an "infrastructural inversion": that is, it takes the "backstage" aspects of examinations and makes them the focus.[2] To paraphrase historian Martha Lampland and sociologist Susan Leigh Star, infrastructure denotes the boring things that exams run on, and like other forms of infrastructure – plumbing, roads, cables – the material objects and repertoires of exams are taken for granted and noticed only when they fail.[3] Their very familiarity and success make them difficult to discern as parts of a larger system.

Historical distance makes it easier to perform an infrastructural inversion. This book studies one period when there arose many now-familiar objects and repertoires of testing, when candidates came to believe that the exam in front of them was a valid test of their knowledge, and that others were being asked exactly the same questions. Its focus is the British Isles between 1850 and 1900, mostly England, as well as glances to the wider empire. In 1850 there were almost no mass achievement exams there. By 1897 their administration was described as one of the era's "new industries" by the chemist and educational reformer Henry Edward Armstrong.[4] Critics of this new system, such as the Oxford Assyriologist Archibald Henry Sayce, warned of an "adopted Chinese culture," ironically just as the power of the famous classical examinations of China was waning.[5]

Previous accounts of Victorian exams, most notably Andrew Warwick's *Masters of Theory* and William Clark's *Academic Charisma and the Origins of the Research University*, have studied the smallest and most famous examples, especially the Mathematical Tripos of Cambridge University. Although both are astonishing works of scholarship to which this work is greatly indebted, both focus on a tiny number of elite and usually male candidates writing at a single location. While this book does discuss the Tripos, it is just as interested in mass exams such as those of the Department of Science and Art (DSA), which between 1861 and 1900 administered over two million tests in fine arts as well as in science, technology, and mathematics. If it is an important research question to ascertain how audiences for science, technology, and mathematics were transformed from elite groups into mass audiences,[6] then there would seem to be few better places to look than at examinations written by millions. Also known as "South Kensington" – for the department's location – the DSA simultaneously tested thousands of candidates in exams ranging from mechanics to animal physiology, fine art, steam engines, and higher-level mathematics. By 1882 over 750,000 students had written DSA exams – an industrial scale.[7] And overseas, by 1885 the DSA model of using examinations to specify science curricula was used in both India and Upper Canada (now Ontario).[8]

It is, of course, anachronistic to say that the Victorians were engaged in "standardized testing": better to say that this book is a history of the *emergence* of the materials and repertoires that made standardized testing possible. What we now call "standardization" was called at the time "securing uniformity," with a single exam written outside a single institution in multiple locations known as an "external examination." This emergence took about fifty years: in 1850, there was little examination infrastructure. Armstrong's 1897 announcement of a new industry across the British Isles and British Empire was a recognition of the existence of this infrastructure. By 1897 the basic repertoire for a mass exam had been established: a specialist in a given topic would draw up questions, which would be identically printed out in large numbers, often in the thousands, shipped to numerous locations, and simultaneously answered by large numbers of candidates under conditions as similar as possible. The candidates' written answers were then returned to the experts, who marked them in accordance with existing knowledge in that subject. Marking consisted of assigning each answer a quantitative score, which could then be aggregated to yield an abstract representation of a candidate's performance on that exam.

Agency

One reason for the appeal of mass examinations is the promise of a more "objective" method of gauging performance.[9] Although the concept of objectivity will be defined more thoroughly below, at this point it can be equated with neutrality: a detachment from individual desires.[10] Exams seem to work because of their capacity to neutrally *measure* achievement. Yet, paradoxically, there is another quality to mass exams that also makes them appealing – they set *targets*. For instance, as a neutral measure, a widely administered exam can help public servants discover how many eight-year-olds in Ontario understand when a rectangle is also a rhombus. But the same exam can also be used in the service of school administrators who want at least 65 per cent of those eight-year-olds to know when a rectangle is also a rhombus. Neutral measure or publicized aspiration, then? Camera, or engine?[11] Exams are noteworthy for their "miscegenation of 'is' and 'ought,'" in historian Keith Hoskin's memorable phrase.[12] This book argues that this paradox arises because it is reflexive human beings, active agents, who are being labelled and examined.

Hoskin's thoughts about exams are strongly influenced by the work of the philosopher Michel Foucault. Foucault famously depicted examinations as part of a larger endeavour to assess, categorize, and normalize human behaviour. In his view, an exam was a mechanism placing individuals in a "field of surveillance" and "network of writing" that captured and fixed them, pinning each person down in their "own particularity." It "disciplined" people – not discipline in the sense of punishment, but rather to shape them and their expectations. This shaping occurred in the service of what Foucault described as a project of "governmentality" – with people coming to adapt themselves to the needs and demands of the state.[13] The Foucauldian approach explains much, such as how the push to "reform" schooling in England aimed not simply to rationalize school organization but also to turn schools into places where examination numbers became meaningful, to paraphrase historian Theodore Porter.[14] In this view the state, and other authorities, wished students and exam candidates to become more legible.

Yet a danger in adopting too strong a Foucauldian approach would be to overlook how examinees themselves responded to all these exams.[15] After all, examinees were, and continue to be, intentional, self-aware agents, not dopes and suckers. This is the very reason why exams can both measure and set targets – because when people have labels put upon them, they tend to change their behaviour. At work is a dialectical pattern. Or an arms race. Thus we will see how examinees changed their own behaviour in response to examination – deploying *counter-repertoires*. Historical distance lets us more easily see the emergence of these counter-repertoires, many of which are still practised today. In addition to simply accommodating themselves to exams,[16] candidates often tried to frustrate, subvert, or play the hard-won standardization of the new mass exam infrastructure to their advantage. One response was to "cram" for exams – to provide a simulacrum of knowledge that would convince an examiner that the required material had indeed been learned. Some Victorian tutors believed that some of their geometry students – despite memorizing the first four books of Euclid's *Elements* – had never understood a single proposition in it. I confess that when I test my own students, sometimes I entertain similar doubts.

Another counter-repertoire to exams was misconduct, now called "cheating." Misconduct meant more than simply stealing glances over someone else's shoulder: it systematically ignored the very rules that ensured uniformity. For what is exam cheating but to act in ways that flout standardization? We will see teachers – financially desperate, or often cynical about the whole purpose of examinations – who obtained the exams ahead of time and drilled their students in the correct answers. They attempted to make their students *less* legible, and such strategies are still followed today. But we also study a more hopeful counter-repertoire – to take seriously the implicit promise that exams tested not only for knowledge but also, indirectly, for moral worthiness. For one reason why mass exams were popular at mid-century was that they were also supposed to reward merits such as industriousness and self-discipline. Reformers of female schooling, then, campaigned to take exactly the same exams as males. They wanted to show that girls and women knew as much as boys and men, and also possessed similar virtues. When women did succeed on such a level grade as the Cambridge Mathematical Tripos, their victories became stubborn facts – and moral claims – that could be used in political struggles such as women's fight for the right to vote.

Examination Jacobinism

"How to learn and what to learn?" asked the Irish visionary educator James Booth in 1856.[17] To try to answer such a question was to be faced with a bewilderingly complex schooling landscape. We will see how in the British Isles at this time there were multiple schooling regimes, sometimes funded by government and sometimes not, shaped by class and gender expectations, and usually fractured by sectarianism. Booth, as architect of another exam system that emerged in the mid-1850s, that of the Society of Arts (not to be confused with that of the DSA), offered a simple answer: sidestep questions about educational *process* and focus on the demonstrated *results* of education. Examinations tested results, reinforcing what historian Sheldon Rothblatt has called "the famous English distinction between teaching and examining."[18] The Society's exams aimed to produce concrete, tangible evidence of the knowledge acquired by working men who had studied alone or at Mechanics' Institutes. Educational reformer Frederick Temple shared Booth's belief in results rather than process, and to this end, in 1856 and 1857 he helped create the Local Examinations of Oxford University. Also in 1856 and 1857, a similar belief in results over process led to the New Charter of the University of London, which focused entirely on examinations to sidestep concerns about the trustworthiness of some of its less-reputable affiliated colleges.

Indeed in 1856–7 appeared a phrase that de-emphasized process in favour of clearly demonstrated outcomes: "payment on results." Its meaning was most succinctly stated by one of its architects, the chemist and public servant Lyon Playfair: "The State wishes a certain knowledge in Science; it is willing to pay for this. The knowledge may be had in the School, College, or garret, by books, oral demonstration, or Experiment – in any way – if its attainment is effected, we will pay for it."[19] At this time, Playfair was a co-head of the DSA, and this principle informed a new policy to pay teachers for student successes on its exams. The intention was to stimulate demand for the study of art, science, technology, and mathematics, as many teachers were struggling to recruit enough students to make a living. By 1859 the DSA had set out a system of "payment on results." The details were worked out by Playfair's DSA colleague, the pioneering civil servant Henry Cole, and his assistant, the Royal Engineer John Donnelly. Three years later, the Education Department adopted the same policy, with the preposition changing to become "payment *by* results." The Education Department, which was responsible for the schooling of mostly poor children at government-funded national schools, based a school's funding on each student's exam passes in reading, writing, and arithmetic; these monies often affected teachers' pay.

Starting in 1857 with the cry for "free trade in education" and ending in 1862 with "payment by results," this five-year period was a time of examination Jacobinism. Beliefs articulated then still seem to shape attitudes today in the British Isles and the former colonies of the British Empire – particularly regarding the economic logic of attending school and possibly even the greater monetary value of studying science, technology, engineering, and mathematics (STEM). Achievement in these subjects was seen as important for economic reasons, not merely because it would improve British industrial competitiveness, but because the DSA's STEM exams paid better.

The focus on results that drove this frenzy of examining was followed by a Thermidorean reaction in pedagogy that emphasized learning processes and underlying comprehension. The "heuristic" movement was one example of this. It was during this period of reaction that scholars wrote the first histories of the DSA and Victorian examinations, and their tone is almost always one of horror. Such an attitude still shapes modern scholars' concerns about monetizing marks. As an educator myself I share this sentiment. But writing as a historian, it has been necessary to sympathetically reconstruct the assumptions underlying policies such as payment by results – something necessary if we are to understand the emergence of mass examinations in the British Isles and thence to other parts of the empire. This is what Part I, "Examinations," discusses.

Examinations, "Meritocracy," and Abstractions

This work contributes to the history of science education and the history of education more generally, yet it uses a more restrictive definition of "education" that is better described as *schooling*: the formal or institutional provision of education. Merely in these two fields it is important to list some debts to historical scholarship appearing over the last twenty years since this project first began: work by Ruth Barton, William Clark, Alex Craik, Janet Delve, Matthew Eddy, Benjamin Elman, Paula Gould, Hal Hansen, Andrea Jacobs, Richard Jarrell, Claire Jones, William J. Reese, Leonard Schwartz, Josep Simon, Christopher Stray, Gillian Sutherland, Andrew Warwick, and Kangmin Zeng.[20] And for works written before 2000, this book owes much to such scholars as William Brock, Harry Butterworth, Carol Dyhouse, Frank Foden, Graeme Gooday, Thomas Kuhn, David Layton, Roy MacLeod, Rita McWilliams-Tullberg, Russell Moseley, Joyce Pedersen, W.W. Rouse Ball, John Roach, Sheldon Rothblatt, and Michael Sadler.[21]

Yet because examinations are not merely tools to assess educational attainment, any history of examinations must go beyond education and schooling. Exams were also used to gauge an individual's fitness for a particular position – for instance, to ascertain the competence of prospective civil servants. Or exams might test someone's worthiness to enter an institution or a profession. In short, exams promised to discover "talent" and foster a "meritocracy." This word, like the term "standardized testing," is also an anachronism[22] – although it does appear in the title of Benjamin Elman's book about the exams of Late Imperial China; he defines meritocracy as a "merit system" that promoted talent.[23] In a similar way, John Carson's comparative history of France and the United States studies how "merit" was conceived and changed from the Enlightenment era to the present day. Carson's conclusion is relevant for this book: since "merit" is so multivalent, meaning such different things to different people, it can only be defined against a larger background of cultural assumptions, knowledge systems, and moral economies.[24] Indeed, as a thought experiment, it is difficult to imagine any social system that *does not* claim to promote some form of "talent" and "merit" among its members. If Elman's expansive definition were strictly followed, all societies could claim to be "meritocracies." This is not to say that "merit" or "talent" does not exist, or that it is incommensurable between cultures, but that there are numerous forms of merit, and so one must define one's terms precisely.

What, then, were mid-Victorian visions of "merit" or "talent"? The proponents of examinations – whether politically Radical or Liberal – almost always depicted virtues in the language of political economy. Examinations induced a healthy competition that rewarded knowledge. To succeed on them, one needed to exercise self-discipline and long-range planning. The highest performers on exams – including elite students of mathematics – demonstrated such "masculine" traits as self-denial, emotional toughness, and physical striving.[25] Examinations would therefore reform schooling and bring about "free trade in education," a phrase reminiscent of the anti-Corn Law campaigns. Above all, examinations would reward the hard work of *individuals.* Such implicit virtues were appealed to whenever promoters of "literary" examinations had to defend them against sceptics who asked how a demonstrated knowledge of Euclid's Proposition Six[26] would help a future cavalry officer lead a charge, or who charged that the entrance exams of the Indian Civil Service created "competition wallahs" suited to a desk and unsuited for their positions in the real world.[27] Against such charges, figures such as Benjamin Jowett and Charles Edward Trevelyan could respond that while exams tested candidates' knowledge and achievement, they also indirectly tested other attributes such as industriousness. In 1854 William Edward Gladstone even assured Queen Victoria that exams for the Indian Civil Service acted as de facto tests for gentlemanliness[28] – unsurprising, since these exams closely resembled those on which Gladstone had excelled at Oxford.

The most direct claim that exams tested both knowledge and morality was made by Thomas Macaulay in 1854:

> Early superiority in science and literature generally indicates the existence of some qualities which are securities against vice – industry, self-denial, a taste for pleasures not sensual, a laudable desire of honourable distinction, a still more laudable desire to obtain the approbation of friends and relations. We therefore believe that the intellectual test which is about to be established [for Indian Civil Service candidates] will be found in practice to be also the best moral test which can be devised.[29]

This quote is given at length because it was repeated at least three times, first by the Civil Service Commissioners in 1858, then in a 1900 description of the Colonial Civil Service, and then again in Robert Moses's 1914 dissertation on the British Civil Service that informed his failed attempt to reform New York City's own civil service.[30] The belief that achievement exams also assess virtues strengthens their capacity to mix "is" and "ought" together, to both measure and set targets. This belief places such exams at the heart of what we think of as a meritocracy.

Moving beyond the question of merit, there is another way in which examinations do not simply belong to the history of education: they yield certificates and other credentials, or what today's economists sometimes call "signals." In 1856 and 1857 the results of exams were championed as ways to replace untrustworthy personal testimonials with trustworthy "mint-marks" or "hallmarks" about a person's character, that is, symbols likened to forms of currency. The reforming civil servant Edwin Chadwick explicitly linked exams to the mathematician Charles Babbage's notion of "saving the labour of verification"; that is, exams saved people from having to spend their own time inquiring into the abilities and character of another person.[31] Seen as tools with which to gauge competence, or talent, or knowledge, exams also served to make individuals (and the larger institutions to which they belonged) legible. Exams could be used to generate an abstract representation of a complex system, be it a school, a school district, or the activity of a candidate's mind.

Examinations also belong to the history of statistics. Historian John Roach showed how examination results came to be analysed with statistical tools,[32] and the use of statistics nowadays to analyse the results of large-scale tests is seen as obvious. But there are steps between mass examinations and the numbers they yield. It is necessary to first understand how these abstract symbols came to be seen as meaningful representations of complex activities. Historian of statistics Alain Desrosières suggests that the "material procedures of objectification" needed to derive meaningful statistics involve three steps: keeping written records; scrutinizing and assembling them according to a predetermined grid, or tabulation; and interpreting them as numbers. The philosopher Bruno Latour has told us to "never speak of 'data' – what is given – but rather of *sublata*, that is, of achievements": how information is shaped to become usable evidence. The historian Jon Agar alerts us to how the form of a government file pre-structures the content therein.[33] Such insights inform Part II of this book, "Examiners": how exams were made into written records, making it far easier to partition questions and answers into separate entities and preset categories. But even before they could yield usable numerical representations of the answers, still another step was required.

By "preset categories" I mean that in order for examination statistics to be seen as meaningful, the categories of comparison – as well as the examination candidates – had to be seen as *commensurable*. People had to be confident not only that what and who were being compared were worth comparing, but that it was even possible to compare them. Meaningful quantification is only possible after commensuration. The sociologists Wendy Nelson Espeland and Mitchell L. Stevens point out that commensuration picks out a common relation among things or properties that seem qualitatively different.[34] It is necessary for these relations of similarity to be recognized before any meaningful quantification can be undertaken. For instance, age is one kind of relation of similarity: no matter how different each individual fourteen-year-old is, they are the same age.

Commensuration is the decision to compare them because of this common property. Hence in 1858, examining a group of fourteen-year-old boys on how well they knew the first four books of Euclid's *Elements* – assigning each answer numerical marks, and ranking each boy by adding up these numbers – was seen as unproblematic. But in that same year, most examiners would have refused to do exactly the same thing for a group of fourteen-year-old *girls*. Such a rejection would have been based on their belief that girls were innately different from boys, and so too should be the goals for their education. This belief in their qualitative difference meant that boys and girls were seen as *incommensurable:* any comparisons between those groups would be unfair or invalid, let alone any quantitative rankings of their performances.[35]

Now, of course, comparing girls and boys is unproblematic. So how did this change come about? How did incommensurability change to commensurability? Liberalism and first-wave feminist calls for equal treatment were important, but so too was the mundane examination infrastructure that had been created: registration, enumeration, standardization, categorization, tabulation. Mass examinations accompanied the Victorian love of statistical collection, which not only included numbering the people but also sorting, categorizing, and labelling them. These activities also involved the same parties – government bureaucracies, institutions, and reforming groups. While many liberal and feminist reformers pushed for boys and girls to be treated as commensurable, another hidden factor was a simple organizational fact: it was more economical to run fewer assessments against fewer standards. It was faster, cheaper, and easier to test twenty people using a single examination than it was to test two sets of ten people using two different exams. We will see that whenever there occurred multiple examinations for different groups of people – such as special "women's exams" separate from men's – there was pressure to cut costs and save time by merging them. It is well known how many women of the time publicly complained about the inherent unfairness of being tested against a separate standard that would be deemed inferior to the one used by males. It is less well known how, privately, examiners who had to create and run these second exams were asking for pay raises because of the extra effort involved. Taken together, these two factors helped effect small changes.

Of course, when candidates were put into categories – first-class pass, "plucked" (i.e., failed), third-class pass – many did not passively accept their placement. This is the central point of Part III, "Examinees." Many worked to subvert this enumeration and these categories with numerous ingenious counter-repertoires: cramming, coaching, or cheating. The effect was to make themselves "illegible" to the institutions seeking to learn about them. Institutions in turn responded with their own attempts to make those people legible again, often by plugging loopholes in the rules– making the larger process resemble an arms race.

Definitions

Because scholarship on exams in my own realm of history and philosophy of science has tended to discuss aptitude exams such as IQ testing,[36] it is worth carefully defining the words this book will use. One is the word "examination." One reason why the topic of examinations goes beyond education and schooling and into issues of competence and potential is the multiple, even paradoxical, nature of examinations. Because exams measure and target, directly testing some qualities while indirectly assessing others, they can be both diagnostic (explaining the past and the present) and prognostic (suggesting the likely future).[37]

Despite the multivalent nature of examinations, this book uses a straightforward definition: *examinations are devices that assess a candidate against one or more standards of achievement.* Standards can be very different. Does a candidate correctly parse a paragraph from Julius Caesar's *Bellum Gallicum*? describe the structure and function of a trachea? or fail to show these abilities? Assessing means sorting people against a preset goal into different classifications. The contemporary work of sociologist Lawrence Busch has been helpful in formulating this definition of an exam,[38] yet this vision of exams as devices sorting against standards also fits what Victorians recognized. Examinations were constantly being discussed in the same breath as educational "standards" – the need to raise them, to ensure they were appropriate, or to compare them with those of other jurisdictions.

Yet the standards that exams such as the DSA's compared each person against were ones of achievement – that is, the *performance* of acquired knowledge and skills. It bears repeating that achievement exams are different from aptitude exams, which are intended to measure some inherent feature or capacity of a person such as IQ (akin to how "a farmer weighs or tests his pigs or his chickens," to use one evocative image).[39] Where aptitude tests are intended to test something *innate*, achievement tests are supposed to test something *acquired*: how to translate a passage from Thucydides, or show one's understanding of long division by correctly dividing 5,839 by 13. Hence achievement exams usually set out explicit standards in advance, giving candidates time to prepare by gaining the necessary knowledge or skill. And, depending upon what standard is used, candidates may perform up to or even beyond that standard – which *also* enables exams to act as both neutral measures of what candidates know and to set targets for what they ought to know.

Such "miscegenation" arises because unlike the assessment of different species, disease-causing agents, or minerals against some classification, the sorting of humans into groups is far more complex. Labelling an entity as a novel coronavirus does not cause that entity to change its behaviour. But labelling a *person* as suffering from a novel coronavirus certainly does. That is because people are intentional and reflexive agents.[40] The same holds for achievement exams. People changed their behaviour as exams sought to label them and put them in a particular classification, each with its own moral resonance. We shall see how examinees pushed to better fit a more desirable category – "senior wrangler" rather than "wooden spoon" – by studying with coaches, learning "tips" that would "pay," or smuggling in crib notes.

The pattern of bureaucratic sorting and reflexive response to it does not seem to have been denoted by a single Victorian word. "Emulation" – denoting the attempt to race others in some contest, with such competition being key to bringing out each individual's best effort – is close. In postrevolutionary France, such "émulation" was celebrated as being generative of the talents and merit that *ought* to replace birth as the grounds of social distinctions.[41] Yet even this word does not quite capture the sense of tactically focusing one's efforts on the explicit requirements while discarding other forms of knowledge as extraneous. When the payment-by-results system of the Revised Code changed funding so that only passes in reading, writing, and arithmetic would yield government grants, many affected schools simply stopped offering classes in "non-paying" subjects such as music and science. Our more sceptical time has better words and phrases to denote how exams themselves shape what is taught. "Teaching to the test" is the best-known phrase. Another is "reactivity," coined by Espeland and Sauder, to denote how an institution changes its behaviour under the pressure of assessment. Reactivity conveys the tension between the "is" and the "ought" of exams – as neutral depictions of a person's knowledge or ability, and as ways to deliberately bring about changes in that knowledge and ability.[42] Victorians were keenly aware of this tension, recognizing that while one goal of mass examination was to discover how much students learned at schools, candidates' scores were published in order to impel poor performers to improve. We see the same forces at work today when magazines and think tanks publish school rankings often derived from standardized tests.

"Looping" is another term for this blurring of assessment and the person assessed, used by philosopher Ian Hacking to denote how individual people react to the labels attached to them. As with the word "reactivity," "looping" depends on reflexivity – someone is sorted into a category, usually one with moral connotations. These moral connotations affect others' actions towards the categorized person as well as the person's own self-understanding, shaping the possible intentional actions he or she can make.[43] Someone asked to divide 5,839 by 13, and who fails to do this correctly, may not only be defined as ignorant in long division; they may come to see themselves as ignorant. This perception in turn sets out the possibilities of what they may do next – seek help from an instructor; leave the class permanently; sit closer to someone else next time and surreptitiously copy their answers. A model Hacking adopted from Foucault to think about the history of statistics can also be used to think about the history of exams. Hacking speaks of two "vectors." One vector, from Foucault's "bio-politics," comes from experts from above who test, classify, and bestow labels on people below them. The other vector, from "anatamo-politics," bubbles up from below, coming from those being tested and being classified: either they accept those labels as new ways to live, or they struggle for autonomy from those labels, thwarting the intentions of the experts.[44] Parts II and III of this book follow, respectively, the downward and upward vectors: Part II surveys the efforts of various examiners to organize and streamline mass exams, and Part III charts the work done by those being tested to alter their classification by those exams.

Returning to our definitions of exams: even limited to a specific time and place, what one calls an achievement exam still covers an enormous amount of territory. It can denote a test to see whether a seven-year-old child of Standard I could add and subtract numbers "of not more than three figures"[45] or a trial of endurance to pass a set of sixteen separate papers held over eight January days on the 1867 Cambridge University Mathematical Tripos. So it is necessary to give still more specific definitions to precisely set out the book's scope.

A Victorian exam could assess against a *pass* standard: a fixed level usually set out in advance, which a candidate either met or did not meet. The arithmetic exam for the seven-year-old child noted above was a pass exam. Or an exam could assess against a *competitive* standard, which compared individual candidates relative to one another, ranking them. The Mathematical Tripos was a competitive exam. The competitive standard was marked by uncertainty: because one did not know how well other candidates might perform, everyone found themselves working harder than they might otherwise prefer. Thus while pass standards were often used to assess competence for a position, competitive standards were used to allocate possession of a limited number of desirable posts such as a fellowship, government job, or other prestigious position. Many Victorian examination systems combined pass and competitive standards, for instance first holding a preliminary "qualifying" (pass) stage, then a second competitive stage.

Ardent reformers such as Edwin Chadwick disliked pass examinations because the standard was explicit, allowing a candidate or coach to gauge the minimum amount of work necessary to scrape through. In our own parlance, pass exams were easier to "game." Reformers favoured competitive standards because they encouraged emulative behaviour. That is, not only would candidates have to work harder to keep up with their rivals; teachers would be forced to work more "efficiently." The intention was for exams to reform institutions like schools and public offices; by being competitively assessed, the standards would rise automatically and mechanically, pulled up as though by an invisible hand. We shall see the affinities between this belief and the political economists' belief in the efficacy of free trade, shown by reformers' push for "free trade in education."

An exam held at a single school, usually in a single form (class), was an *internal* examination. A test of candidates from different schools was an *external* examination. By the late 1850s external exams had been coordinated and standardized to the point that the same questions could be answered by candidates at more than one location, simultaneously, and after the session the answer booklets were returned to a central location and marked all together. Most of the exams discussed herein were external, though not all of them.

What about the material objects used in an exam? This book focuses mostly on paper-based exams. Questions were printed in advance, and each candidate furnished their answers by writing them out on paper. Obviously, tools such as pens, ink, paper, and writing-sand to dry wet ink were used, but written exams required other things too – a writing surface of some sort, a chair, a quiet location in which to write, a way to mark time elapsed. As exams became more uniform, so too did the conditions in which examinees expected to work. When these expectations were not met – a construction crew noisily digging outside, for instance – then complaints followed about this, rooted in a concern about lack of uniformity, causing unfairness.

A written exam can be contrasted with a practical exam and an oral exam. The use of oral exams – also known as *viva voce* or "living voice" – declined through the century, although they continued to be used at such schools as Oxford. Sometimes these were also called "public" examinations since they were carried out in front of an audience. Practical exams, meanwhile, came especially out of chemistry, the fine arts, medicine, and botany, following the insistence that to show knowledge of these fields it was necessary to demonstrate certain skills such as the identification of parts or specimens. Practical exams became important as a way to supplement written exams and to prevent memorization without understanding – that is, *cramming*. T.H. Huxley's new 1872 system of teaching biology, using his innovative teaching laboratory at South Kensington, emerged not simply out of his desire to improve science teaching: it was also intended as a defence against cramming.

Victorian exams could be paper, practical, or oral, or sometimes combinations of them. As the century progressed, the paper format became dominant. The shift was mainly due to administrative considerations. Practical and oral exams placed greater strains on examiners because such tests required them to be present when the questions were asked. In the 1830s oral examinations at the Royal College of Surgeons ran from dawn until midnight. By contrast, paper exams partitioned the question and answer, and the questioner and answerer, from their immediate contexts – thus identical questions could be delivered to multiple locations; answer booklets could be taken away and compared with one another. In 1846 examiners testing pupil-teacher candidates for the Education Department first seem to have realized that they did not have to be present at the exam itself. This permitted the spread of such delegated "proctored" paper exams to faraway places, and by 1863 this practice had become global.

Oral and practical exams also had difficulty maintaining their perceived uniformity and impartiality, particularly when taken by large numbers of candidates. Candidates tended to share exam questions with others. So if an examiner wished to ask the same question of each candidate, one had to go to great lengths to keep all candidates separate, going so far as to imprison them in waiting rooms for hours. By contrast, paper exams made possible the simultaneous asking and answering of questions. They were easier and less expensive to run. Such considerations encouraged the use of paper-based exams over oral and practical ones.

Relevant Questions

This book addresses a number of questions currently of interest to historians of science and the wider scholarly community.

First, if the history of science is in danger of becoming ever-more local, taking contextualization of a given episode as a worthy goal in itself, then how does one move beyond such parochialism to something more general, while still acknowledging the importance of local settings? One suggestion is to study patterns in the circulation of knowledge.[46] As noted above, this book depicts examinations as one highly successful case of knowledge in transit, spreading to places like India, Trinidad, and Mauritius as well as the rest of the British Empire.[47]

If we focus on the British Isles, there was a relatively small and genteel audience for science in the first half of the nineteenth century. By the second half of the century, the audience was becoming larger – a "mass" audience. How and why?[48] This is another question asked in this book. To be sure, it was important that books became cheaper and that new forms of mass communication emerged, along with an army of science popularizers.[49] But this book suggests that scholars of science popularization have overlooked the industrial-scale science examinations not only of the DSA but also of the Society of Arts, the College of Preceptors, and the City and Guilds.[50] In the "marketplace for attention" that was the British Isles, such exams created a high demand for books on the syllabus, as it was only by studying those books that one could pass the exams. In this light the most committed and attentive reader of one of T.H. Huxley's writings was likely someone studying for his exam. Where historians such as Jonathan Topham have asked the subtle question about how practitioners of science rhetorically constructed new audiences for science,[51] this book focuses on a far less elegant process: getting *paid* for receiving high marks on tests. Moreover, the explicit link between doing well on a science examination and getting paid for it – an appeal to the "cash nexus" – would certainly have given the study of science a more instrumental image.

Third, how did a "man of science" make enough money to do science, especially when in the 1850s serious practitioners were expected to write articles for which they did not receive payment?[52] Examining was an important way for such people to pay the bills, and the number of men of science who acted as examiners is long and surprising. For instance there was a chance that someone writing the DSA's physiography (physical geography) exam would have the famed biogeographer Alfred Russel Wallace – the co-discoverer of evolution by natural selection – marking their answers.[53] Indeed, acting as an examiner was something that united numerous scholars of the period, no matter which of the two cultures they belonged to. The poet Gerard Manley Hopkins was an examiner in Classics for the Royal University of Ireland and marked hundreds of papers. Another poet and school inspector, Matthew Arnold, marked exams for national schools as he sat at the deathbed of his youngest son, Basil.[54] Such mundane and time-consuming activities tend to go unnoticed even in the best historical works – for instance Adrian Desmond's splendid, all-encompassing biography of Huxley never mentions that each May, this extraordinarily busy man had to spend three weeks to organize the marking of zoology and animal physiology exams for the DSA, exams that numbered in the thousands and then tens of thousands.[55]

Fourth, how did there emerge in the UK a "scientific priesthood" committed to promulgating scientific naturalism in the population and to policing the meaning of scientific terms? Desmond has focused on the role of what Grant Allen called Huxley's "drilled and dragooned South Kensington student,"[56] but in order to even get to study in Huxley's South Kensington laboratory, one needed to first score highly on the DSA exams he set. Numerous members of the X Club, a group founded to reform the governance of British science, were examiners not only for the DSA but for various other institutions. They often specified textbooks they themselves had authored. Indeed, one might even say that early on the DSA was "captured" by key members of the X Club like Huxley and physicist John Tyndall.

Fifth, how did the standardization of exams imprison some and emancipate others? To make things commensurable, diverse possible forms of life and knowledge must be reduced, abstracted, made "flatter" or even homogeneous; such impoverished routines can be imposed upon unwilling individuals. Yet on the other hand standardization can reduce misunderstandings and confusion; it can promote rules; it can reduce the arbitrary actions of powerful people who would otherwise have too much discretion.[57] This book emphasizes the Janus-faced character of "standardized" Victorian exams. The oppressive nature of standardized exams has been discussed at length, often by neo-Foucauldians who like many scholars today see formality in social life only in terms of its pathologies.[58] Yet the standardization of exams often made it possible for people seen as qualitatively different to *ask* to be viewed as commensurate with people in a different group. The point is noted when we look at women's campaigns to be compared using exactly the same exams that men took, under exactly the same conditions, such as the Cambridge Mathematical Tripos. These women even celebrated exactly the same moral traits, such as competitiveness and industry, that exams were supposed to reward as forms of merit.

Most abstractly, it has been shown that texts are polysemic – imbued with different meanings, from which an individual can select and create an amalgam of knowledge.[59] This raises the sixth question of how it was possible for larger epistemic communities to emerge. If individuals do take different meanings from texts, then how do groups of people come to generally agree upon the same facts and definitions?[60] One answer is that it was examiners who then (as now) pushed back against candidates' "polysemic" interpretations of their textbooks. Examiners could shape an epistemic community – say, all those required to correctly understand the anatomy and physiology of the trachea – by policing interpretations, marking some as correct and others as incorrect. Such enforcement seems to be part of the Janus face of standardization – that is, to set and mark an exam is to restrict alternative definitions, to promote epistemological conformity, even orthodoxy. Yet at the same time such conformity is necessary for common knowledge in a community – not just knowing something, but being aware that *others* comprehend roughly the same thing by the same word. Such common knowledge facilitates coordination and makes scientific knowledge truly social.

Organization of the Book

This work ranges across a number of different subjects that do not always immediately relate to science and technology, and sometimes not even to schooling or exams. This approach is necessary to establish examinations as part of a much larger setting inhabited by educationists, their institutions, and their intentions. One cannot understand the DSA's policy of "payment on results" without also understanding that group's rivalry with the Education Department, for instance. This book has also extensively used witness testimony at various parliamentary commissions, which gives us a more detailed view of how examinations actually worked (or failed to). What results is a picture of an "ecosystem" of examinations in which the same people pop up in different contexts. For convenience, Appendix B provides brief biographies of many of these people, and Appendix A gives the major dates of events covered in this book.

Part I: Examinations lays out histories of exams, with successive chapters moving from a wide perspective to a very narrow one. Chapter 1 is a history of the "mania"[61] for examinations across the British Isles and wider British Empire, focusing on the second half of the nineteenth century. To better orient the reader, it investigates exams in Classics, the sciences, fine arts, and mathematics. It argues that the standards against which candidates were assessed were infused with additional moral values that helped people associate achievement exams with merit. Chapter 2 is a history specifically of the exams of the Department of Science and Art, and then the Education Department's 1862 Revised Code for elementary education. The two are linked because it was the creators of DSA exams who went on to design much of the Revised Code. The chapter thus explores how exam results were taken to be sufficient evidence not only of students' knowledge but also of teachers' "efficiency." Chapter 3 is even more fine-grained, taking up the workings of the DSA's animal physiology exam of May 1873, with the questions drawn up by Huxley and answered by the candidate Charles Ledger. This chapter is strongly informed by historiographic attention to how the "mundane" practices and "ink-stained fingers" of invisible labourers make possible high-level and abstract activities.[62] It then moves on to think about how examinations were used to enforce some degree of conformity in order to make epistemic communities possible.

Part II: Examiners takes the examiners' perspective, focusing on how they created and managed Victorian exams. It likens examiners both to authors and to collectors of statistics. Chapter 4 looks at how exams became increasingly written down on paper, and what this shift made possible. One change that emerged was the specification of increasingly precise standards to which a candidate performed. Another was the greater ability to partition candidates' answers from their immediate temporal and geographical context, making it easier to compare answers. Still another was the possibility of ranking candidates ever more finely. The chapter relates exams to the period's fascination with statistical collection and to what Foucault called a "field of surveillance" that situated each candidate in a "network of writing."[63]

Chapter 5 focuses on what the Victorians called "securing uniformity" – setting out standards, and deciding which particular groups or individuals were worth comparing (in other words, who was deemed *commensurable* against a given standard). Then it looks at how exams were standardized, and if enough uniformity was attained, quantifying marks on answers to better compare them. Chapter 6 is about exams, publicity, competition, and what Theodore Porter has called "thin description" – that is, decontextualized, abstracted labels that allow a person to more easily describe a complex system.[64] The credentials resulting from exams became signals. Exams sorted people into categories if they possessed a particular quality, thereby giving them a label and membership in a certain group. For such efforts, examiners are called anti-anthropologists.[65]

Part III: Examinees takes the examinees' perspective, likening them both to audiences for a work and to people being classified. It focuses on examinees' agency. These three chapters look at the ingenious ways in which candidates responded to exams: acceptance to or resistance of the standardization that went into exam processes, and the thin descriptions that the exams were supposed to impose upon them. Chapter 7 looks at how examinees changed their behaviour to meet or surpass the standard, either by studying well or by improving their ability at exam writing. As educationists distinguished between *content* and *form,* so too did examinees. "Crammed" knowledge, in this light, was the obverse of "thin description": one might call it "thin knowledge." The chapter charts not only rising standards but also a new industry created by exams – science tutors such as Herbert George Wells who, before he became a world-famous author, taught zoology students how to defeat the examination machine. Chapter 8 looks at how some examinees went even further to manipulate this classification process. In modern parlance it looks at how examinees *cheated* on examinations, charting the various tricks and dodges that made it appear that they had successfully performed to the required standard. The chapter's key case is the Goffin Affair of 1878, in which a successful science teacher was discovered to have been opening up DSA exams in advance and teaching his students the answers to the questions immediately before they took the exams. He did not take seriously the implicit morality of examinations.

Finally, Chapter 9 studies people who *did* take this morality seriously, not only changing their own behaviour but trying to change the attitudes and behaviour of those seeking to classify them. It looks at how marginalized groups took the efforts made to standardize exams – and the moral values infusing them – and turned them back against the examiners. Excluded groups did so as part of a larger claim that they *should* be seen as commensurate with other groups of people. The specific case discussed is a campaign by reformers of female schooling, reformers who petitioned for girls and women to be tested on exactly the same exams as boys and men. Although the most famous representative of this campaign was Philippa Fawcett in her 1890 victory on the Maths Tripos, the real architect of the campaign was Emily Davies, founder of Girton College, Cambridge, who insisted that the same standards ought to be applied to both women and men, even when those entrenched standards were loathed by many as antiquated. Foucault showed how examinations made it possible for teachers and other authority figures to conduct "surveillance" of students;[66] by simplifying performances, exams *did* make it easier for such "top down" assessments. But "bottom up" scrutiny could also happen. Written examinations facilitated "sousveillance," the assessment of authority figures' actions, comparing them with the rules they claimed to follow and the virtues they championed, and pointing out when their actions didn't measure up. It is by shaming Goliath that Davies sometimes wins.

PART ONE

Examinations

1 "The Age of Examinations": A Historical Sketch

I have said that the extended use of the instrument of examinations is eminently characteristic of the age in which we live. I would almost venture to say that, amidst all the material and all the social changes by which the period has been distinguished, there have been few that are greater or more peculiar than this ... let us write among its titles this – that, as it is the age of humane and liberal laws – the age of extended franchises – the age of warmer loyalty and more firmly established order – the age of free trade – the age of steam and railways – so it is likewise even if last and least, the age of examinations.

– William Ewart Gladstone, Prize Ceremony, Association of Mechanics' Institutes of Lancashire and Cheshire, 23 April 1862

This chapter is a broad overview of the Victorian examination mania between 1846 and 1895. It is intended for the reader who may not know, or who may wish to review, key details of the period. It discusses such famous exams as the Cambridge Mathematical Tripos, as well as lesser-known ones such as those of Trinity College Dublin, the University of London, and the College of Preceptors. It then discusses the widespread adoption of examinations in the 1850s, a growing disillusionment with them by the 1870s, and their condemnation in an enormous 1889–90 petition. Key dates can be found in Appendix A . One goal of the chapter is to show how different exam systems were themselves parts of a much larger exam ecosystem. During this history we will see the same organizing figures popping up in different places – men such as James Booth, Harry Chester, T.H. Huxley, Benjamin Jowett, and Frederick Temple. Many are listed in Appendix B. Their involvement in several examination systems helps explain why a new repertoire created in one system soon found its way to other systems.

Exams classified candidates against an explicit standard, such as a particular level of knowledge. Further, exams were seen as instruments to encourage talent since they also tested for less-explicit moral criteria: desirable traits such as industriousness and self-discipline, for instance. In May 1854, writing to support the examination of civil servants, philosopher John Stuart Mill argued that although no test could directly assess "moral qualities," exams did so indirectly: a "well cultivated intellect" was usually accompanied by "prudence, temperance, and justice, and generally by the virtues which are of importance in our intercourse with others." The twinning of explicit standards and implicit virtues made exams seem as if the bookish qualities they were testing for were not arbitrarily selected.[1]

Education, or Schooling?

Before proceeding further it is important to distinguish schooling from education. Books with such titles as *The Case against Education* show how the two concepts can be confused,[2] but the historian Gillian Sutherland wisely insists on the difference. Since she uses Emile Durkheim's definition of education as the "methodical socialisation of the young," any educational process, in addition to school, would include networks of family and friends, as well as independent reading, jobs, or hobbies.[3] This book uses Durkheim's definition of education while adding a more traditional rider that denotes learning as a form of acquisition: learning *how* (skills) and learning *what* (knowledge). Schooling, then, can be seen as a specialized version of education, in that it is often methodical and often involves socialization, and learning certainly occurs in schools. But simply going to school does not automatically lead to an education, as though osmosis is at work.

Another reason to insist on the distinction between schooling and education is the role played by credentials. Schooling seems to be best able to confer trusted credentials that denote the *possession* of education; today, credentials are sometimes called "signals." But these can only be trustworthy when others bestow them. If I declare myself to be highly educated, without evidence from others, it would – and should – be difficult to believe me. As formal institutional places of learning, schools allow other people, like teachers, to assess levels of knowledge and skill through continuous scrutiny and recording of achievements. So Foucault taught us.[4]

In the Victorian British Isles the distinction between education and schooling was insisted upon by many. Her Majesty's School Inspector (HMI) John Norris, who surveyed government-funded national schools for poor children, contrasted the "indigenous" system of education in England with the system of "public instruction" on the Continent.

While public instruction was instituted by central governments, he preferred the English system because of its spontaneity, and he hoped that English schools would continue to be "established by the people themselves."[5]

For a comparative perspective, in 1868 Charles Cassal – an examiner in French language and literature at the University of London and a graduate of the French education system – testified to the Taunton Commission that although the French system was inexpensive and mixed together students of different social classes, it was structured on military lines and tended to undermine the personal initiative of teachers. He held the English system to be superior because, paradoxically, its very incompleteness forced students to be self-reliant.[6]

As a former member of the National Assembly, Cassal confirmed the famous anecdote that the-then French minister of public instruction (M. Duruy) would glance at a clock and claim to know what was being taught in every French school at that moment, such was the uniformity of the French schooling system. In contrast, Cassal saw and praised the extraordinary diversity in the multiple schooling systems in England, as well as the parochial schools of Scotland and those governed by Ireland's own Board of Education.[7] A central reason for all of the different school systems was sectarianism: an effort by the Church of England to structure primary schools led Dissenters to establish their own system out of reluctance to send their children to a school run by a different religious group. One can nonetheless find exceptions: in Leighlinbridge, Ireland, the future physicist John Tyndall, son of an ardent Protestant Orangeman, was inspired to a life of science and scholarship by the gifted Catholic schoolmaster John Conwill. And a school inspector observed numerous Jewish and Catholic students attending Anglican primary schools in Bristol, where he claimed that "the religious difficulty, as it is called does not exist."[8]

What about girls and women? Further disunity of school systems emerged from the view that female and male students ought to be kept separate as they were headed for separate spheres[9] – thus schools were largely intended for boys only, with only a few for girls. There were a few exceptions, where girls attended the same schools as boys, notably government-funded national schools. Yet until the late 1850s when discussions about schooling began in earnest, be it by parliamentary commissions or well-meaning pamphleteers, "schools" almost always meant ones intended for boys. Even the most liberal-minded reformers sometimes used the word "boys" when they meant both male and female children. This forgetful ignorance was cleverly seized upon by reformers of female schooling to get the topic of girls' schools included in the mandate of the Taunton Commission of 1867–8, given that its mandate had been set out as "education given in schools not comprised within her Majesty's two former Commissions."[10] The former commissions were the Newcastle Commission of 1861 and the Clarendon Commission of 1864: the Newcastle was dedicated to a study of the conditions of government-funded national schools, and the Clarendon to the elite public schools.

That it took three different parliamentary inquiries to understand how schools were managed points to the central reason for the fragmentation of English primary and secondary schooling: class and social status. This can be contrasted with the more egalitarian Scottish parochial schools. In England when parents did opt to send their child to a school, they not only chose schools where their child would socialize with members of the same social class but also actively avoided schools where "rough children" were numerous.[11] So strong was this fear that even where a school for poorer children was acknowledged to be instructionally superior, parents still avoided it. In 1874 the future biology exam coach H.G. Wells was sent to the badly organized Morley's Academy rather than the well-regarded national school closer to his home in Bromley.[12] Government funding signified the presence of "rough children" in a school. For poor families struggling to survive, fees might be too high (schooling was always supposed to cost something), so parents might pull their children out of school to earn money for the family.

Indeed, there were further school groupings only partially covered by one of the three commissions. At the very bottom of the social scale were workhouse schools and those known as pauper or "ragged" schools: some lessons were taught there, like adding up sums of money and reading from scripture. One inspector noted that the main intention of these schools was to "correct the roving propensities" of children and improve their character; often such children were identified as candidates for emigration to the colonies.[13]

For poor families aspiring to slightly higher social status, the "national school" was organized by either the Church of England or Dissenters. National schools were intended for children of tradesmen and the working poor. Before 1839 these schools were established only after private money had been obtained through school fees or neighbourhood subscriptions; "free" schooling was always looked down on. Yet such payments were usually not enough to offset the cost of running a school. After 1839 and some agitation, the government set aside funds to pay for school construction and teacher salaries. Government money was given on condition that the school be inspected once a year and that a school's teacher "certificated" – that is, certified as having passed a training regime. The role of the "monitor" – an older child, good at their lessons, who helped teach the younger children – evolved in 1846 to something more like a formal apprenticeship. These students were designated as "pupil-teachers," receiving training from the certificated teacher and some money from the government.[14] Pupil-teachers, too, took exams and stayed in the school for longer periods. But they were the exception: children attending these schools tended to join the workforce in order to help support their families, and so they usually remained at a school for only a few years.

Moving upward in social status, proud they received no government money, stood schools intended for middle-class boys. There was no consistent definition of such institutions: for numerous reasons related to funding, history, or organization, they were known variously as "grammar" or "private" or "private adventure" schools. Even what all of these schools seemed to have in common – their attendance by middle-class students – was disputed. Sometimes a school's managers denied it was "middle-class" because such a description contradicted their aspirations towards gentility.[15] At any rate, because middle-class schools received no government money, they required no certificated teachers or inspections. As a result the organization and procedures of such schools varied. By the early 1860s some prepared students for commerce; others for university or military academies; a tiny number were technical and scientific. Students at middle-class schools tended to leave around age sixteen, usually going into the workforce, with a few headed to university. There were also a few schools for middle-class girls, but those that did exist emphasized class distinctions too: while the North London Collegiate School for Girls aspired to higher social status and charged higher tuition fees, its sister school, the Camden School, was less expensive yet offered less chance for social advancement.

Of the highest status, and also receiving no government monies, were "public schools" such as Eton or Winchester. This group was formalized by the Clarendon Commission after it studied nine such institutions. At these places, students tended to remain for longer periods, often leaving at age eighteen for university. Such schools sought to cultivate a social elite in habits of gentility and leadership, especially through the study of Classics. Some saw the Classics as useless or irrelevant, but this irrelevance was entirely the point. While Classics did offer a common set of cultural references, it mostly denoted social status. Children struggling through Classics often did so because their parents insisted that they learn what Classics they could – Latin at a minimum and Greek if possible.[16]

Numerous excellent studies of the schooling landscape can be found in the references, and for more information the reader is pointed to those. The point of this sketch of the numerous schools and their systems (where any existed) is to show what underlay this disunity: the assumption that people from different groups were *qualitatively distinct* from one another, having not only different needs and future trajectories (e.g., "separate spheres") but also different innate capacities for learning. Different schools prepared students for very different forms of life. This institutional disunity encouraged a belief in the qualitative distinctness and the incommensurability of different kinds of students, a belief that came full circle when defenders used it to explain away why all the different school systems were so disunified. Conversely, what various "reformers" of schooling held in common was a belief that students from various realms might not be *that* qualitatively different, and that they could be seen as commensurate and therefore meaningfully compared. This root belief informed much agitation in the 1850s for common and explicit schooling standards that facilitated comparisons between different groups. While places like France had a centralized ministry to standardize what students learned, "pushing" the curriculum upon them, in the British Isles, examinations came to act in the opposite way, seeking to "pull" numerous different schools and their students *towards* the standards for which they tested.[17]

Alexis de Tocqueville argued that amidst the chaos of pre-Revolutionary France, with its "traditional, confused, and irregular" constitution, separate ranks, and unequal burdens, people came to imagine a new society that stood for the opposite of this confusion: in this imaginary new society, "everything appeared simple and coordinated, equitable, and in accord with reason."[18] Historian Ben Kafka points out how powerful to French citizens this vision of the state was, even if that aspiration never quite lived up to reality. Examinations offered a similar vision for many school reformers in the British Isles: in a chaotic and fragmented schooling landscape, the power of exams lay in the hope that they could show whether someone knew something or not.

Some Exam Systems

Cambridge University

The mathematician Augustus de Morgan called Cambridge University the "parent of the *examination-paper.*"[19] It was also the parent of the belief that competition was the feature that made examinations work. The history of exams at Cambridge has been comprehensively told – not just recently by William Clark, Alex Craik, John Gascoigne, Christopher Stray, and Andrew Warwick, but also in earlier accounts by W.W. Rouse Ball and Michael Sadler.[20] To briefly summarize, in 1730 Cambridge instituted the Senate House Honours Examinations. Students seeking a BA were tested either by viva voce or by disputation. By 1763 candidates were examined in three groups according to merit: "hard-reading," "reading," and "non-reading" men. Respectively, these groups became wranglers, optimes, and poll-men. Candidates in the first two groups came to be ranked individually.[21] By the 1770s the number of questions had increased so that the exams ran for two and a half days; by 1802 it was five. The exams also increasingly used paper, first with the questions dictated, then printed in advance. By 1790 the two largest colleges (Trinity and St John's) set their own exams to assess their own students before the Senate House Honours Exam. By 1792 letter grades in a series of upper- and lower-case letters were starting to be assigned to answers on each question. The practice of giving a numerical mark to each exam answer does not seem to have appeared at Cambridge until either 1820 or 1836, although there are claims this began earlier.[22]

Mathematics was the most important subject in the Senate House Honours Exam. When the exam ran over five days by 1802, mathematics questions occupied the first three, followed by a day of moral philosophy, then a final day of questions for wranglers.[23] When in 1822 the Classics Tripos was instituted, the Senate House Honours Exam became known as the Mathematical Tripos. Proponents of mathematics argued that it trained students in logic and abstract reasoning; moreover, the subject was well suited to examinations because the strict rules governing mathematical operations leave far less room for examiner interpretation – that is, subjectivity.[24] The rigidity of these rules, and the competition between the candidates, helped contribute to the notion that the exam and its rankings were true indicators of talent and merit, free of personal judgment and other fallacies. The Mathematical Tripos became an exemplar of what is now called mechanical objectivity: "the insistent drive to repress the wilful intervention" of a person "through a strict protocol," to use the definition of historians of science Lorraine Daston and Peter Galison.[25] And because Cambridge produced graduates who went on to found or shape other institutions, their upbringing in a system of rigid rules, and the belief that it was unproblematic to use competitive exams to sort people, went along with them. When the new University of Toronto wished to hire its first mathematics professor in 1842, it simply went through the list of that year's wranglers, finally hiring the sixth.[26] Hence when such Cambridge male graduates embarked upon their own careers, often as educators themselves, they often fell back upon familiar repertoires. For instance the historian and Member of Parliament T.B. Macaulay, who wrote the Classics Tripos at Trinity College, thought it reasonable to use exams to assess the potential competence of civil servants; numerous other MPs who were also Cambridge graduates shared his belief. They even had a shared language of ranking: as Macaulay spoke in Parliament in favour of competitive examinations for the civil service, he noted that for every one junior optime who had distinguished himself, there were twenty wranglers.[27] He did not bother explaining what these terms meant, assuming that his fellow parliamentarians would know.

The Mathematical Tripos was an attractive model for other institutions because of additional moral qualities that had come to be associated with it. As Master of Trinity Hall and eighteenth wrangler of 1845, Henry Latham must have had such qualities in mind when he noted how exams called out traits necessary for success in life, such as "enduring hardness" and industriousness.[28] Andrew Warwick has discussed how these and other masculine virtues came to be associated with the Tripos. Above all, this "hardness" emerged out of the Tripos's use of a competitive standard to rank individual candidates. Such virtues were widely publicized – by 1825 *The Times* was printing the rankings of wranglers and optimes. William Clark notes that such publicity invested the wrangler with academic charisma, which in turn increased the appeal of the Cambridge examination system still further.[29] Such competition helped mathematical culture flourish at Cambridge. For instance, the Mathematical Tripos had a tradition of giving an exceptional answer to a question higher than full marks, then taking that answer as the model for future versions of that question. Such incremental improvements raised the level of mathematical achievement, helping turn Cambridge into the world's leading centre of mathematics by the end of the nineteenth century.[30]

Yet far less has been written about the third group of Cambridge students who opted *not* to take honours – more numerous than wranglers and optimes, since the name "poll-men" was taken from the Greek word *hoi polloi*, "the many." These "non-reading" men still had to take some examinations – Sheldon Rothblatt has pointed out that one reason for this was to encourage industriousness while deterring immoral conduct. More time spent preparing for an exam meant less time drinking, engaging sex workers, fighting with townsfolk, or rioting.[31] However, in the first half of the nineteenth century, "Poll" exams were often empty and even cynical rituals. Even in mid-century it was difficult to fail a non-honours exam: in 1841 one eyewitness noted how little was asked of the candidates during one exam. Not only did the candidates sit closely to one another, frequently copying from each other; the presiding examiner often left the room and even seemed to avoid watching them. This conduct was deliberate: as the examiner collected their answers, he remarked that it was the students' fault if they did not pass, "for he didn't turn his back upon them the whole time for nothing."[32]

Trinity College Dublin and Oxford

In 1636 Archbishop Laud instituted a system of "public examinations" at Oxford and Trinity College Dublin, where three masters would ask oral questions on grammar, logic, geometry, and Greek. Although this test soon become a mere formality at Oxford, at Trinity College Dublin exams continued to be taken seriously. Few candidates failed the initial entrance examination for Trinity College; the intention was to fail candidates in later tests. Various prizes and medals were given out for exam performances, with competitive exams starting in 1731.[33] We will see how one Trinity College Dublin graduate, James Booth, advocated for widespread examinations in 1847 and then went on to design the exams of the Society of Arts.

At Oxford, meanwhile, Laud's public examinations had been predated by a 1588 statute requiring each candidate to pass an exam in grammar and logic, and to recite the Articles of Faith and Religion, referring to scripture in support. The 1852 commission inquiring into the state of Oxford University concluded that these early exams had declined because there were no prizes, no candidates were ever rejected, and examiners ("moderators") found it difficult to rank candidates, especially when their performances were relatively close.[34] The commissioners believed that the system began to change in 1800, when a new statute instituted new exams and mandated that examiners' positions be paid ones. This 1800 statute also called for the twelve highest-performing candidates to be publicly listed in ranked order. More changes followed: in 1807, the exams were split into separate subjects, with mathematics and physics separated from the "Literae Humaniores" (Classics, logic, rhetoric, and moral philosophy). Soon thereafter separate examiners conducted the mathematics and physics exams. Then candidates on the basis of their exam performances came to be arranged into four honours groups (first, second, third, and fourth class). However, any way to rank them more finely was made difficult by the practice of marking each exam with Greek letters rather than numbers. Historian Christopher Stray believes this was done deliberately to avoid "intensive ranking," that is, ranking every candidate. Chapter 5 explores the matter of ranking candidates in more detail. While Oxford is famous for using the viva voce format to pose questions, this system placed great demands on examiners' time, with complaints that they were examining over half of the year. This pressure was partly relieved when printed exams were introduced there in the late 1820s.[35]

The 1852 Oxford Commission called for even more exams. Its report strengthened one Oxford faction who had long equated exams with university reform. At its forefront was the classicist Benjamin Jowett, at Balliol College. Jowett would soon be caricatured in Anthony Trollope's 1858 *The Three Clerks* as "Mr. Jobbles," who "had divided the adult British male world into classes and sub-classes, and could tell at a moment's notice how long it would take him to examine them all. His soul panted for the work. Every man should, he thought, be made to pass through some 'go.'"[36] Although he did not become Master of Balliol until 1870, by 1852 Jowett was at the centre of a network of former Balliol students who went on to use or promote examinations. These men included Stafford Northcote (Sir Warwick Westend in *The Three Clerks*), who by feats of prodigious memory had earned a Balliol Fellowship, and who in 1853 called for exams to screen candidates for the civil service; Frederick Temple, a scholarship student who as an HMI ran a teacher training school, Kneller Hall, and who went on to formulate examination systems for the College of Preceptors and the Oxford Local Examinations; Ralph Lingen, who replaced James Kay-Shuttleworth as Secretary of the Education Department, and so brought continuity as various political masters came and went. Arguably Jowett's best-known student was Matthew Arnold, who as an HMI came to detest the "mechanical" nature of examinations. We will return to all of these men later.

The two most important Oxford graduates to promote examinations came from other colleges. At Christ Church, William Gladstone had famously taken a "double starred First" and saw examination triumphs as self-evidently prophesying future worldly success. He became Liberal MP for Oxford, and as a rising political star – eventually becoming Chancellor of the Exchequer, then Prime Minister – he strongly favoured the use of examinations. In addition to his former private secretary and political ally Northcote, Gladstone was helped by Robert Lowe, former Fellow at Magdalene College and Liberal MP. Lowe's vision of education was unsentimental, because for seven years after graduation he was a successful private tutor, or "crammer," in Classics, working ten hours a day with ten Oxford students. Further shaping Lowe's view of the world as a site of competition was his own upbringing – his father was a workhouse-championing reverend – and

Lowe was proud of his own ability to overcome his physical challenges of albinism and near blindness. Lowe accordingly favoured any system emphasizing the value of individual effort – he saw his own career trajectory as a testimonial.[37]

In 1852, not yet an MP, Lowe sent a letter to the Oxford Commission emphasizing exams as part of a larger crusade for "free trade in education." More exams would force a student to work so that he "should never feel himself free from this stimulus"; college tutors ought to be replaced by undergraduates' free choice of private tutors. And since Oxford's college system acted as "clogs to the efficiency of the University," anyone with a Master of Arts ought to be able to open his own Hall. The ensuing "competition from without" would reform Oxford.[38] Upon becoming elected MP, Lowe championed exams as Vice President of the Board of Trade, then later, as Vice President of the Committee of the Privy Council on Education. When Gladstone became Prime Minister, Lowe became Chancellor of the Exchequer, carrying out tasks both large (instituting examinations for the entire civil service) and small (getting Jowett the Balliol Mastership he so craved). Meanwhile, an open secret among his Westminster rivals and allies was Lowe's position as a leader-writer, or editorialist, for *The Times*, a position he used to call for free trade in education. For his effect on examinations and public perceptions of them, Lowe is the most important figure in this book.

East India College, Haileybury

To train future staff, the East India Company founded its college at Haileybury in 1809. There were no exams. One became a student there by being nominated by a company director. Graduation was inevitable after four terms in two years unless one was expelled, but this rarely happened; instead punishment consisted of "rustication," in which the student was not allowed back into Haileybury for a term. Rothblatt's point that exams were often used to ensure student discipline is nicely demonstrated by case of the East India College. After several serious riots – one involving gunpowder – exams were introduced at Haileybury. They were not welcomed – one early history of Haileybury described them as "extorted" from the directors "like drops of blood." First came an exam on "Oriental" subjects in 1813, then on "European" ones in 1820. In 1821 an exam for each term was instituted. By 1840 assessment had been systematized: monthly exams, attendance checks, and frequent scrutiny of notebooks.[39] Student performances were woven into a Foucauldian web of written records.

Haileybury's adoption of such records bears intriguing similarities with the "Cornwallis" push for anglicization and utilitarianism in the East India Company's administration of India. The Cornwallis system emphasized census taking, the methodical collection of other forms of information, and above all, clear rules that were written down. Such a system was championed by Thomas Macaulay in his 1835 *Minute on Indian Education*,[40] and its centre of gravity was Bengal. Another advocate of the Cornwallis system was Charles Trevelyan, who became one of Haileybury's highest-performing students; in 1824–5 he won various prizes and medals in history and political economy, and he led the class in all four terms in Classics and Sanskrit.[41] Upon graduation he went to Bengal; in January 1840 Trevelyan became Assistant Secretary to the Treasury, holding this position until 1859. At Treasury, Trevelyan caught the eye of Gladstone, who in March 1853 commissioned the inquiry known later as the Northcote-Trevelyan report.[42]

Still more strands in a gigantic examination network were woven. As Macaulay's brother-in-law, Trevelyan helped him write the 1854 *Minute* calling for candidates applying for employment in the civil service to be examined, to be discussed below. He came to strongly favour this principle after both the 1848 revolutions and his coordination of famine relief in Ireland. In the first case, Trevelyan saw the 1848 uprisings as a signal that governments had better improve their administrative work; in the second, although he callously tended to attribute the famine to providence, he did think that government relief efforts had been hindered by the incompetence of numerous civil servants.[43] Meanwhile at home, his wife, Hannah Macaulay Trevelyan (sister of Thomas), delighted in calling her husband "Sir Gregory Hardlines" – his alter ego in Trollope's *The Three Clerks*.[44]

Various Military Academies

Examinations and competition have thus far been mostly discussed in the context of *schooling*. But they were also used to sort between those permitted to advance to more senior posts, gain a job, or enter an institution – and those who were not so permitted. In the Royal Navy, for instance, as early as 1677 someone wishing to become a lieutenant had to have served at least three years at sea plus pass an exam in seamanship run by senior officers. Although only about one in twenty failed, such an exam still worked to force all aspiring officers to gain the necessary three years' experience.[45]

Meanwhile, the army had three academies that gradually took on more exams. There was Woolwich, which trained Royal Engineers; Addiscombe, the East India Company's military academy; and Sandhurst, which trained officers. Since 1774 Woolwich required that its candidates know some arithmetic and Latin grammar. In 1809 Addiscombe copied this requirement. In 1812 Sandhurst followed suit, allegedly when the Duke of Wellington was shocked by one cavalry officer's ignorance.[46] Before the 1853–6 Crimean War, however, most exams at these military academies tended to be public, and so were rehearsed performances for audiences. At Addiscombe, for instance, the June 1851 public exam culminated with the performance of the best student of that year: "The questions put to him relate of course to astronomy and Newton's *Principia*. And with ... some eye to effect, this performer delays the production of his board for a while, so as to give the impression that his mind is battling with questions concerning the movements of the heavenly bodies. But once embarked, it exceeds the young man's skill to make it appear that his thoughts are occupied in the solution of an original problem, when they are, in fact, engaged in endeavouring to recall the terms of a well-conned lesson."[47] It was unclear what such exams actually assessed.

A widespread perception of military incompetence fostered by the Crimean War, combined with parliamentary inquiries such as an 1855 Commission on Sandhurst and the 1857 Yolland Report, led to greater scrutiny of potential officers. Instituting stricter entrance exams was one simple option, and in 1857 the new Council of Military Education duly reformed these exams.[48] It was felt that they gave some evidence of a candidate's suitability, despite objections from those who felt the link was not obvious – how precisely would an officer's extensive knowledge of Classics help him lead a cavalry charge? By 1867 the (civilian) mathematician and head examiner for the Council of Military Education, Henry Moseley, testified – possibly a little too conveniently – that before Sandhurst's entrance exams were made more demanding, candidates displayed an "astounding" level of ignorance and "sluggishness of mind." No longer. A change of the Woolwich entrance standard from pass to competitive he believed had driven up the proportion of Woolwich candidates able to perform elementary mathematics from 25 per cent in 1855 to 80 per cent twelve years later.[49] Moseley left unstated a likely corollary: by 1867 the average Woolwich candidate had probably already taken similar exams, and so had become more comfortable with writing out their knowledge on paper than most 1855 candidates had been. After all, by the late 1860s, military exams were occurring at least seventy-five days a year.[50]

Examinations of Aspiring Medics (Apothecaries, Surgeons, Physicians)

To become a licensed *medic* (a word used here to simplify the very different occupations of apothecary, surgeon, and physician) in the British Isles, one had to pass examinations. But there were numerous places where one could be examined, all with different levels of difficulty and comprehensiveness. Such diversity was frequently complained about and exploited: students seeking the easiest exam to take to become a medic often went to the University of St Andrew's, the Society of Apothecaries, or the Royal College of Surgeons. Over time the variety of such exams would be reduced, especially with the creation of the General Medical Council examinations in 1858.

How candidates were examined also differed by format. The exam of the Royal College of Surgeons was entirely oral until 1860. By contrast, the exams of London University, later the University of London, were mostly written. Their questions were usually published afterward to facilitate public scrutiny and foster greater trust. The perception that the Royal College of Surgeons was corrupt, while London was at the vanguard of reform,[51] was intimately tied to the preference of whether things were written down or not. The matter of papyrocentrism will be revisited in Chapter 4 – for now we turn to London's university.

London University – University of London (University College and King's College)

Unlike Oxford and Cambridge, which both allowed entry or degree-taking only to Anglicans, London University was intended to be open to all who lived in the British Isles regardless of class or denomination. Prospective students were assumed to be male, an assumption that women would eventually challenge. London University emphasized the role of written examinations from the outset. Not only did written exams raise the "efficacy" of the lectures – as a new institution it emphasized the "strictness" of its exams,[52] doubtless to compensate for its lack of prestige relative to Oxford and Cambridge. By 1836, when it was reorganized as the University of London – becoming an examining board for University College and King's College – a memo to the Privy Council noted that its exams would be explicitly modelled upon the Cambridge Tripos.[53]

The best-known exam at the University of London was its Matriculation, or entrance, examination. It was strict, with only about half of the candidates passing: in 1838 there were "papers," or separate tests, in seven individual subjects: Greek or Latin, French or German, English grammar, history and geography, logic, moral philosophy, and mathematics (which included arithmetic, algebra, geometry, mechanics, hydrostatics, heat, electricity, optics, and astronomy). Each paper had to be passed. Even more papers were planned: on asking about the Matriculation exam in 1842, a young T.H. Huxley was informed by the university registrar that future exams might include papers in chemistry, zoology, or botany.[54]

The Matriculation exam seems to have been the first exam used in assessment-at-a-distance, at least in the English language,[55] especially because of its extensive use of printed questions rather than viva voce ones. In 1846 the newly founded University of Sydney, in Australia, wished to use the Matriculation exam to assess its own students. This was because Sydney deemed that exam to be "*of great use* even in the colonies. It fixes a standard – not left to the discretion of Professors and Masters, but to the wisdom of a learned body in England." Rather than ship the Sydney students all the way to London to be examined, the request was for the Matriculation to be held in Sydney, with the printed questions shipped there and the answers returned to London for marking. This request was accepted by the University of London's senate in April.[56] The use of the Matriculation to fix "a standard" not subject to the judgments of Sydney's own professors shows the University of London's increasing recognition as an examining institution.

The Education Department (Committee of the Privy Council on Education) before the 1862 Revised Code

In 1834, £20,000 was voted by Parliament to pay for the construction of schools; in 1839 a Committee of the Privy Council[57] was created to distribute these funds, and more money was set aside to establish teacher training colleges and subsidize teacher salaries. The committee's Secretary was James Kay-Shuttleworth.[58] To qualify for some of this government money, what became known as a national school had to be available for inspection. Always university educated, often a well-regarded scholar, in 1839 Her Majesty's Inspector (HMI) for the Committee of the Privy Council on Education was simply directed to "collect facts and information" and report back and not interfere with the running of the school.[59] Over time, additional requirements were placed on funded schools – for instance the requirement that they employ certificated teachers who had trained for three years at a "normal" training college and then spent two years apprenticing.

After entrance and scholarship exams, training colleges first examined each student orally once a week, but this was found to be too time-consuming. So the examination format changed to a one-hour test each day. Aspiring teachers would write out on paper their answers to questions put on a blackboard, ranging from arithmetic to Bible knowledge. These answers were then sent to topic specialists, who gave each answer a numerical mark. When aggregated, at the end of the month the numbers yielded an average – a "convenient number," said Kay-Shuttleworth – representing each aspiring teacher's intellectual progress. Moral qualities were also recorded.[60] Upon gaining a certificate teachers could continue to write exams to earn further certificates in a graduated system, which could increase their salary by £10 to £30 a year.[61]

In 1846 the Committee of the Privy Council on Education created the position of the "pupil-teacher," a teaching apprentice who was selected from among the most proficient students at a national school; taking extra lessons in the evening after helping teach the daily class, each pupil-teacher earned £10 a year in government money, rising to £20 by the end of the five-year apprenticeship. To earn this money they took yearly exams. After this apprenticeship, if the pupil-teacher wished to become certificated as a teacher, he or she could write a national examination to gain admittance to one of the teacher-training colleges, possibly winning a Queen's Scholarship to pay the tuition. By 1849 some 4,000 pupil-teachers were being tested each year, with exam questions published after completion to convey the "standard required" to earn a Queen's Scholarship; naturally, pupil-teachers formed groups to study these past questions.[62]

The 1846 Queen's Scholarship exam seems to have been the first one written simultaneously at multiple centres – a proctored examination. Its procedures became models for other groups' proctored exams. In 1854, as the Society of Arts was devising its own exam system, procedures for the Queen's Scholarship exam were described in the Society's *Journal* on the grounds that they were not yet "generally known."

> The papers of the candidates in all parts of the country are sent to the Central Office where they are sorted according to subjects, and sent to different inspectors, e.g., all the papers in Arithmetic to one, all in History to another, and so forth. Each inspector assigns a number of marks to each papers, according to its merit. The papers are then returned; those of each candidate are put together again; the total number of marks which he has obtained is ascertained; and the candidates are finally arranged according to the result of the comparison.[63]

Notice here the administrative ease of the routine. Efficiency was obtained by using topic specialists, marking the exams quantitatively, and, as with the University of Sydney's use of London's Matriculation exam, moving the papers (rather than the candidates) from central office to local centre and back.

The most famous product of such a training system was the fictitious Mr. M'Choakumchild of Charles Dickens's *Hard Times* (1854), who with 140 other schoolmasters had been turned in the same factory as "so many pianoforte legs." At that training college, M'Choakumchild had learned "Orthography, etymology, syntax, and prosody, biography, astronomy, geography, and general cosmography, the sciences of compound proportion, algebra, land surveying and levelling, vocal music, and drawing from models" – and was moving onto mathematics, the physical sciences, French, German, Latin, and Greek.[64] Dickens's contemporaries did agree that certificated national school teachers seemed to know a lot, but they were more sympathetic than he was to them. At one prize exam for schoolmasters and schoolmistresses in Hampshire and Dorsetshire, William Benjamin Carpenter – brother of Mary Carpenter the school reformer, who was himself a physiologist and the University of London's Registrar – proclaimed himself to be "quite astonished at the exactness, and, I may say, perfection of the knowledge up to a certain point" the candidates showed on their papers. He believed that the organization of the national schools ought to be imitated, if only middle-class grammar schools would set aside their snobbishness towards their purported social inferiors.[65]

The College of Preceptors

By eschewing government money, "middle-class schools" did not have to be inspected; although many desperately needed funds, their refusal was mostly founded in their desire to remain distinct from lower-status national schools. At a time when the schoolmaster was often caricatured as incompetent or malicious – exemplified by the figures of Squeers or Wakford in Dickens's *Life and Adventures of Nicholas Nickleby* (1839) – many non-certificated teachers wanted to raise the public status of teaching. They also sought to learn from one another ways to better teach their students. One result was the 1846 establishment of the College of Preceptors. Two of its founders, James Wharton and Richard Wilson, were both Cambridge graduates who had written the Mathematical Tripos. In 1847 the college started examining teachers in the theory and practice of education; twenty-four passed and received a diploma.[66] The college also introduced mathematics exams for middle-school teachers; presumably the routines followed would have been similar to those of the Cambridge Mathematical Tripos, since the examiners included former high wranglers such as John Hind and James Joseph Sylvester.[67]

Financial desperation spurred the College of Preceptors to make the poor decision of giving college membership to anyone paying the required fee. This harmed its reputation. But its financial desperation also led to the decision to start testing the *pupils* of any college member who requested it. This improvisation was far more successful: starting in 1850, by 1851 girls were being examined alongside boys, and by 1853 the college was testing twenty schools.[68] Its exams became an organizational model for the 1857 Oxford Local Examinations, which along with Cambridge's soon overshadowed those of the College of Preceptors. Nonetheless, the college's exams remained quietly successful. Certificates for passing the College of Preceptors' exams came to be accepted in lieu of other certificates for professional qualifications – by 1865 a First Class result could be used to bypass the qualifying exams of the General Medical Council as well as those of the Incorporated Law Society. By 1879 the college was examining 7,000 students a year from 700 schools, about the same number as those taking Cambridge's Local Examinations.[69]

The Society of Arts

After the Great Exhibition of 1851, Mechanics' Institutions were unified under the umbrella of the Society of Arts. Here were run numerous classes for adults – John Tyndall, for instance, learned much chemistry, mathematics, and physics at the Preston Mechanics' Institution – but by 1851 their membership numbers were in decline. Examinations were proposed to reverse this. James Booth had already proposed national exams in his 1847 *Examination the Province of the State*, which argued that successful candidates could use the certificates they gained as testimonials for potential employers. Booth joined the Society in 1852, became a member of its council along with Henry Cole and Harry Chester, and in December 1853 helped plan its first examination system. This first exam, held March 1854, was a failure, with only a single candidate appearing; but by its third iteration in June 1856, it ran both in London and in Huddersfield, testing sixty-two candidates. Frederick Temple was one of the examiners.[70]

The Department of Science and Art

For the modern reader, the Society of Arts (now the Royal Society of Arts) is easy to confuse with the Department of Science and Art

(DSA) – indeed, one official testified that many Victorians were similarly confused.[71] But the two groups were quite different, especially since the DSA was a government department. As the DSA will be thoroughly explored in the next chapter, at this point some connections in the social network will suffice: in 1854 both of the DSA secretaries, Henry Cole and Lyon Playfair, were also members of the Society of Arts, with Cole a long-time member of its governing council. It is thus unsurprising that, like Temple, they too would imitate some of the Society's examining repertoires. Meanwhile, both Playfair and Cole corresponded extensively with others about exams, particularly Chester and Lingen over in the Education Department, as well as other reformers such as Edwin Chadwick. The DSA contact with Chester and Lingen deepened when in 1856 the Department of Science and Art was placed, together with the Education Department, under the governance of the Committee of the Privy Council on Education.

The Civil Service

Chadwick first called for examinations to "reform" the civil service in an 1829 article detailing the French *concours* medical examination.[72] Indeed, unmentioned thus far is how often the French *concours* was pointed to as a model worth imitating, along with the belief in the benefits of a little competitive émulation. His piece advocated a similar system for England and Wales; later, Trevelyan and John Stuart Mill both cited Chadwick's article as moving them to favour civil service exams.[73]

Trevelyan and his brother-in-law Macaulay, along with Northcote and Gladstone, agitated for the use of exams to select civil servants. Gladstone, as Chancellor of the Exchequer, commissioned the Northcote-Trevelyan Inquiry. In 1853 the India Act allowed appointments to the Indian Civil Service to be made by competitive examination; Lowe, as MP and Secretary of the India Board, was informed by Jowett about how these exams might be structured. In November the Northcote-Trevelyan Report argued that examinations should be used to make appointments to all parts of the civil service.

Propaganda for civil service exams emphasized not only that testing would raise efficiency among civil servants, but that it was feasible. Trevelyan was assured that the exams of Cambridge and the University of London had shown it would be "easy" to examine 400 candidates at a time, using neutral examiners, printed papers, and a numerical scale in order to ascertain the "relative merit of examinees."[74] The system used by the Education Department's pupil-teacher exams would be used to ensure speedy marking and "uniformity of standard": exams would be divided up into topics, or "papers," and completed answer scripts sent to the relevant examiner-specialist to mark.[75] The first civil service exams were run in May 1855, with 309 candidates failing out of 687.[76] These *rejections* – not the passes – were seen as the truly valuable results, showing that not just anybody could join. At first such exams were used sparingly, assessing only candidates for the Treasury, the War Office, and the National Debt Office. They were also gamed. Since exams for a position required a minimum of three candidates, to ensure a specific applicant would get hired, in the late 1850s the Treasury kept on hand a "special couple known as the 'Treasury Idiots'" who could "barely read or write" and used them as competitors against the favoured candidate.[77] It was only in June 1870 that competitive examinations were imposed by Gladstone, Lowe, and Lingen on the entire civil service.[78]

Supporters of the exam argued that it indirectly tested various desirable moral attributes. The educationist Richard Dawes spoke of how such tests rewarded perseverance and deterred idleness. He even likened competitive examinations to free trade – just as with free trade, the competition of an exam was virtuous and purifying for its candidates.[79] Queen Victoria was reassured by Gladstone that an open examination was "also an effectual test of character" – proficiency on it could only be obtained by industry and self-denial, which could be "rarely separated from general habits of virtue."[80] He believed that competitive examinations would reward gentlemen and "the aristocracy of this country … who may be called gentlemen by birth and training," because of their "*insensible* education, irrespective of booklearning."[81] To adopt Christopher Stray's point, Gladstone believed new civil servants would demonstrate their intellectual abilities by examination and their respectability by attendance at elite public schools.[82]

Merit meant not only talent and knowledge but also gentlemanliness. Yet not all people agreed with Gladstone's sanguine view. Trollope, who also worked at the Post Office and was thus a civil servant himself, viewed the exams with bitterness as they would only reward boys most effectively crammed by tutors to answer "strings of questions." But above all Trollope believed that some positions could only "be filled by gentlemen." While such a social distinction existed in reality, he claimed, competitive examinations were "based on a supposition that there is no difference."[83] Both Trollope and Gladstone were concerned about which groups could be legitimately assessed and which could not. Who could be commensurated? Trollope's insistence on only "gentlemen" being suited for the civil service is one such concern about illegitimate comparison.

It is worth noting that such concerns about commensurability did not extend to young non-white men: it was deemed unproblematic to compare them with young white ones. Only eight years after the civil service exams began, men of South Asian ancestry began voyaging to London to write it. In 1863 Satyendranath Tagore of Calcutta (Kolkata) placed forty-third on the exam and joined the Indian Civil Service as a sessions judge.[84] He was followed from South Asia by other civil service candidates, who often studied from twelve to sixteen hours a day, since they had travelled to a different land and culture to compete against the natives. But such candidates were often unsuccessful because one British legal convention differed from their own. In India, when someone attested they were eighteen, it meant that on their birthday they had *completed* their eighteenth year of life. In Britain (where a birthday denoted the *start* of a new year of life), that same person would be counted as seventeen. The Civil Service Commissioners were unaware of this difference, meaning that throughout the 1860s they wrongly thought "Hindoo" candidates were older by a year than they actually were. In 1868, using this British convention to misinterpret the ages given on candidates' certificates, the commissioners ruled that four South Asian candidates were over twenty-one and therefore too old to write the exams.[85] The problem stemmed from a lack of awareness that there could even exist other forms of accounting for age – an injustice emerging out of imperfect categorization.

The Oxford Local Examinations

In early 1857, schooling reformers and Oxford scholars seeking to change middle-class schools started their own experimental examination. It is unclear why they decided to venture into a jurisdiction already claimed by the College of Preceptors and to a lesser extent by the Society of Arts, although they were certainly already aware of these exams – some, like Temple, even being examiners.[86] Different social class is the likely explanation. At any rate, Temple was inspired by the notion that natural talent ought to be recognized, and that any system of education ought to discover it. After all, he himself had been a poor young man so "discovered," with his education subsequently sponsored by benefactors, prizes, and fellowships.[87]

In the spring of 1857 the Oxford Local examiners drew up a syllabus in subjects from Classics to arithmetic, languages to chemistry, and sent it to various schools in the Exeter area, announcing they would test knowledge in those subjects. On the morning of Tuesday June 16th, 107 fourteen- and sixteen-year-old boys showed up at an Exeter hotel, paid a fee, received an identifying number, and inside standardized booklets wrote out answers to printed questions. The room's layout is shown on this book's cover. A neighbourhood committee of respected citizens administered this exam, and thereafter formally vouched for the candidates' good conduct on it. After three days of writing, the completed answer booklets were sent back to Oxford for the examiners to grade. Answers deemed good received high numerical scores; answers deemed bad received low or no scores; these numbers were aggregated into certain categories, and the top-scoring candidates received prizes. All those who passed were also given non-degree certificates with Oxford's imprimatur: Associate-in-Arts.[88]

Much triumphalist myth-making about the Oxford Local Examinations immediately followed. Members of both the Society of Arts and the College of Preceptors were irritated, and rightly concerned, that the Oxford Locals would lure away potential candidates because of Oxford's greater prestige. And they complained that none of the Oxford examiners had ever taught at the secondary level.[89]Although pedagogically reasonable, this complaint missed the larger point about the purpose of exams: the examiners' lack of teaching experience was intentional. For the separation of teaching roles from examining roles was what made the Oxford Local exams seem trustworthy. An analogy was to the judicial system and its drive to avoid conflicts of interest and any perception of such conflicts.

The Oxford Locals combined numerous repertoires from other exam systems into a single infrastructure. They were advertised as devices to assess the "efficiency" of teaching at middle-class schools. Soon after they were written, at least twelve petitions from all over England were sent to the organizers of the Oxford Locals asking that their students write these exams for the next year, 1858.[90] Indeed, a key innovation on the Oxford examiners' part was to allow regional groups to voluntarily form "local boards" that administered the exams in each district; the respectability of each board member served to guarantee good exam conduct.[91] (Society of Arts officials retorted that they first had the idea in 1854.) In June 1858 the Oxford Locals were written at ten different centres throughout England, with 800 junior and 423 senior candidates. This was an elevenfold increase in the number of candidates from the previous year. The top-ranking candidates' names were published in order of merit in *The Times*, an image shown in Figure 1.[92]

Examinations such as the Oxford Local became more prominent as they tested ever more candidates. As their scale increased, it became less expensive to test each additional candidate, since the underlying infrastructure was already in place. But more importantly, as the number of candidates taking a given exam grew, the exam and its requirements

THE MIDDLE CLASS EXAMINATIONS.

OXFORD, Aug. 4.

The following results of the "Examination of Persons not Members of the University," which was held in the week commencing the 21st of June at Oxford, London, Bath, Bedford, Birmingham, Cheltenham, Exeter, Leeds, Liverpool, Manchester, and Southampton, has been this day issued by the Delegates. We publish the names of those candidates only who attained some distinction :—

SENIOR CANDIDATES.

SECTION A. (ENGLISH).—FIRST DIVISION IN ORDER OF MERIT.

Name.	School.
Clarke, E. G. ...	... Crosby-hall Evening Classes
Clifford, C. W. ...	... Proprietary S., Edgbaston
Marquis, J. ...	... Clare Mount S., Wallasey
Fell, H.	... Grammar S., Appleby
Moule, H. C. G....	... Fordington Vicarage
Drewett, W. H....	... Grammar S., Burton-on-Trent
Stocks, S. H. ...	... Proprietary S., Huddersfield
Graham, H. J. ...	... Grammar S., Appleby
Rawle, H. J. ...	... Mansion-house S., St. David's, Exeter
Lethbridge, E. ...	... Wellington-villa, Mannamead
Swanwick, J. A.	... Proprietary S., Edgbaston
Evanson, C. P. ...	... Grammar S., Bristol
Long, T.	... Clapham-park S.
Bigham, J. C. ...	... Dorotheenstaedtishe Real-S., Berlin.

SECTION A. (ENGLISH).—SECOND DIVISION, IN ALPHABETICAL ORDER.

Name.	School.
Atkinson, G. T....	... Grammar S., Appleby
Beale, J. S. ...	... Proprietary S., Edgbaston
Bell, J. R. ...	... Collegiate Institution, Liverpool
Benecke, E. C. ...	... Collegiate S., Camberwell
Bott, A.	... Collegiate S., Camberwell
Candolo, H. S. V. de	... Cotham, Bristol
Chapman, V. W....	... Avenham-house S.
Cooke, J.	... Park S., Birkenhead
Croft, A. W. ...	... Wellington-villa, Mannamead
Crosthwaite, W. H.	... Grammar S., Leeds
Dangar, J. G. ...	... North London Collegiate S.
Devonshire, H. J.	... King Edward's S., Birmingham
Drew, W. ...	... Probus S.
Ford, T.	... Bluecoat S., Glocester
Gandy, C. ...	... Windermere College

JUNIOR CANDIDATES.

FIRST DIVISION, IN ORDER OF MERIT.

Name.	School.
Reynolds, R. J. ...	... Windermere College
Gardner, W. R. ...	... Carlton-terrace S., Liverpool
Clifford, W. K. ...	... Mansion-house S., St. David's, Exeter
Richardson, W. F.	... Windermere College
Pullibank, J. ...	... Grammar S., Kingsbridge
Coon, J.	... Craufurd College, Maidenhead
Wilkins, A. S. ...	... Collegiate S., Bishop's Stortford
Appleby, A. ...	... New Kingswood S., Bath
Beaufoy, W. S. ...	... Grammar S., Sutton Valence
Page, R. ...	... Chatham House, Ramsgate
Roberts, Griffith	... Normal College, Swansea
Prest, H. E. ...	... New Kingswood S. Bath
Blacket, E. R. ...	... Woodspean Academy, Newbury
Evans, H. H. ...	... Grammar S., Bristol
Bennett, H. ...	... Collegiate S., Brixton-hill
Webb, J. E. ...	... Huddersfield College
Warr, H. D. ...	... Royal Institution S., Liverpool
Watts, B. H. ...	... Rectory Middle S., Bath
Legg, J. W. ...	... Diocesan S., Portsea
Bytheway, H. ...	... New Kingswood S., Bath
Creed, J. A. ...	... Great-house, Newton Abbot
Evans, H. F. ...	... King Edward's S., Bath
Smith, G. A. ...	... Grammar S., Newton-le-Willows
Harrison, T. W....	... Grammar S., Denmark-hill
Waymouth, S. ...	... Tor S., Torquay
Massy, V. T. ...	... The Hermitage, Bath
Genge, E. H. ...	... The Hermitage, Bath
Obbard, A. N. ...	... Collegiate S., Camberwell
Cross, John ...	... Nelson House, Devonport
Bethell, H. ...	... Cotham
Balls, T., ...	... N. London Coll. S., Camden-town

SECOND DIVISION, IN ALPHABETICAL ORDER.

Name.	School.
Adrian, A. W. H.	... St. Thos. Charterhouse, Goswell-street
Adrian, A. D. ...	... N. Ln. Coll. S., Camden-town
Andrew, A. ...	... Daisy-bank House, Manchester
Baines, E. M. ...	... Mansion Grammar S., Leatherhead
Bayliffe, A. M. ...	... Academy, Chippenham
Bean, A. H. ...	... Royal Grammar S., Colchester
Beavan, A. ...	... Grammar S., Bristol
Betts, G. L. ...	... Cmmrcl. S., Bury St. Edmunds
Bird, A.	... Craufurd C., Maidenhead
Bowden, F. J. ...	... Collegiate S., Greenwich
Buckell, F. J. ...	... Queenwood C., Stockbridge
Bures, E. G. ...	... Denmark-hill Grammar S.

SPORTING INTELLIGENCE.

BRIGHTON RACES.—WEDNESDAY.

The BRISTOL PLATE (handicap) of 50 sovs. One mile.

Mr. T. Brown's Hebe, by Herbalist, 3 yrs, 6st. (C. Brown) 1

Count Batthyany's The Courier, 3 yrs, 6st. 6lb. (Bradley) 2

Mr. Barber's Abron, 3 yrs, 8st. (Dales) 3

Mr. Fry's Cantrip, 4 yrs, 6st. 6lb. (Hiley) ... 4

Count Lagrange's Trouvere, 5 yrs, 8st. 5lb. (Plumb) ... 0

Mr. Simpson's Magnet, 4 yrs, 6st. 10lb. (Walley) ... 0

Mr. King's Naughty Boy, 4 yrs, 6st. 10lb. (Custance) ... 0

Mr. Montague's colt by Weatherbit—Dagobert's dam 3 yrs, 5st. 10lb. (W. Bottom) 0

Betting.—3 to 1 against The Courier, 4 to 1 agst Hebe, 5 to 1 agst Abron, 8 to 1 agst Trouvere, 10 to 1 agst Cantrip.

Won by a length; two lengths between the second and third; three lengths between the third and fourth; Naughty Boy and Magnet were next, some distance from Cantrip.

MATCH for 100 sovs. h. ft. Bristol mile.

Mr. R. Ten Broeck's Orianda, by The Cossack, 4 yrs, 7st. 11lb. (Fordham) 1

Lord Clifden's Indulgence, 5 yrs, 8st. 7lb. (A. Day) ... 2

Even betting.

Orianda made the running, and beat Indulgence, who was not persevered with towards the finish, by 10 lengths.

THE BRIGHTON NURSERY STAKES (handicap) of 10 sovs. each, with 100 added. Winners extra. T.Y.C (six furlongs).

Mr. Barratt's Electric, by Fallow Buck, 8st. 7lb. (J. Goater) 1

Mr. Howard's Flitch, 7st. 7lb. (Fordham)... 2

Mr. Gratwicke's Nassau, 7st. (Cresswell) 3

Mr. J. Barnard's Pan, 7st. 7lb. (Swift) 4

Mr. J. Barnard's Conceit, 6st. 7lb. (Reeves) 0

Duke of Beaufort's The Elk, 6st. 7lb. (Bray) 0

Mr. T. Bell's Eltham Beauty, 6st. 6lb., (Pritchard) ...

Mr. Rogers's Otterbourne, 6st. 4lb. (Custance) 0

Count Batthyany's Northern Night, 6st. (Hiley) ... 0

Mr. J. Evans's Clara Webster, 5st. 12lb. (A. Edwards)... 0

Mr. Saxon's Vigo, 5st. 10lb. (Cooper) 0

Mr. Payne's filly by Cotherstone—Catalpa (W. Bottom) 0

Mr. A. Williams's Palm Leaf, 7st. 10lb. (F. Adams) ... 0

Mr. Harcourt's Julie, 7st. 2lb. (Walley) 0

Betting.—3 to 1 agst Electric, 4 to 1 agst Clara Webster, 10 to 1 agst Flitch, 10 to 1 agst Nassau, and 100 to 8 agst any other.

Figure 1. "The Middle Class Examinations" and "Brighton Races," *The Times*, 5 August 1858, 10.

solidified in the public eye as a de facto common standard with which to compare people. Such a perception began to attract candidates from ever farther away.[93] By the early 1860s, only a few years after the Oxford Local Examinations began, Scottish teachers were sending pupils to England to write them; in response, Scottish and Irish universities such as Edinburgh and the Queen's University in Ireland defensively created their own Local exams.[94]

Exams in the British Isles: 1858

In a speech demanding that girls and women be allowed to write Local exams, school inspector John Norris described the organizing system of the Oxford Locals "one of the great discoveries of our day" because of its combination of "economy of labour; absolute uniformity of standard, and all the security of the utmost publicity."[95] Although Oxford examiners like Temple clearly did not discover any of these underlying features, the Oxford Local Examinations did codify and routinize the repertoires that, when combined, made possible what we can now recognize as "standardized testing." It is worth analysing Norris's three points to better understand what he meant.

Economy of labour meant that paper-based exams were easier to administer. Because there were no viva voce exams, except in specific cases[96] examiners didn't have to be present when the exam was taken. Instead, they delegated its administration to others. Since the answers were written down on paper, examiners had the time to look at the responses afterwards, which facilitated comparisons. The division of labour meant that topic specialists designing the questions and marking the answers only had to pay attention in their own subject of expertise. Such a focus meant exams in specialized topics were easier to mark. Meanwhile, adding new candidates, or groups of them, was relatively easy. In its use of the division of labour, parallels with the factories and manufacturing systems that made possible the industrial revolution are easy to discern.

Absolute uniformity of standard – standardization was secured by a number of factors. The printed questions were identical. The paper answers could be taken away, out of the immediate context of the exam centre and candidate; those answers could be compared with one another at the examiner's convenience. More subtly, other tools of standardization were used, such as rules to ensure uniform writing conditions at each exam centre. Not only did these include a lack of noise and distraction, but also minutiae such as a prescribed candidate distance from each other. But there was an accompanying requirement for *boundaries* on who could be commensurate. One example is grouping by age. On the Oxford Locals, boys up to age fourteen wrote one exam, the "junior"; boys between fourteen and sixteen wrote a different exam, the "senior." Junior and senior exams were not to be compared because each group was deemed incommensurable. Gender also meant incommensurability. It was assumed by many that girls and boys were qualitatively distinct. No girls applied to write this particular exam, and had any done so they would have been turned away. This assumption was not universal, of course – Norris himself was publicly calling for girls to be examined with boys. He believed, therefore, that individuals of either gender *were* commensurable. Finally, boundaries about *what* subjects were to be compared depended upon a particular exam system. The Oxford examiners insisted that candidates be sorted by performance on specific subjects, while the Cambridge Local exams (discussed below) lumped together and ranked numerically all performances. This yielded a single mark that could then rank candidates. To use different categories and not mix them, as in the case of the Oxford examiners, is to assert the qualitative difference of the different subjects: to be skilled at arithmetic is not the same as being knowledgeable in Classics. By contrast, to lump together the resulting marks from all subjects and aggregate them into a single score, as in the case of the Cambridge examiners, is to assert the commensurability of the different subjects.

Security of the utmost publicity was obtained by publicizing as much as possible about the exams. Just how much was actually revealed differed by exam: we have seen how the exam syllabus was published in advance and how questions were often published after it was completed. The identity of the examiners was usually published. As the "league table" in Figure 1 shows, not only were candidate names and results shown in ranked order, but their school or tutor was also identified. Such publicity was thought to generate trust in the exam and draw attention to the most effective teachers; teachers could in turn increasingly be judged by outsiders such as parents or government officials, even if those outsiders didn't quite understand what a good education really meant.[97] Such publicity also enhanced examinations' miscegenation of is and ought – both testing standards and prescribing desirable targets.

The year 1858 was key in the standardization of testing because by then, proctored exams had become normalized. The repertoire of conducting exams entirely out of sight of the very people who had drawn up the questions had become unproblematic to many. With no limit on how many questions could be printed, it was possible to add new exam centres ad infinitum (and the cost for each candidate declined as the number of candidates increased). As long as each organizing committee ensured that writing conditions were kept as similar as possible to those at other centres, simply by following rules already laid down, then it was possible to treat the resulting answer booklets as commensurable. Two articles written twenty years later noted that this period marked the height of popularity of exams in England – a belief that institutions could be reformed by putting positions in them to "a sort of intellectual auction," allowing "men of the largest mental wealth" to come to the forefront and serve their country.[98]

The various exams discussed above make up only a partial list of all the different examinations that emerged in the 1850s. There were numerous others, both internal and external, using either a competitive or a pass standard. In addition to exams for professions such as law or medicine, there were other qualifying tests – for instance those of the London-based Institute and Faculty of Actuaries (1855) and the Edinburgh Actuaries (1856). Both used the same repertoire as the Mathematical Tripos.[99] Meanwhile, entrance exams for specific schools were instituted, or made more demanding. When Temple became Headmaster of Rugby School in 1858, he instituted a competitive entrance exam that he thought produced a "clear and perceptible difference" in the preparedness of students. This was mainly because the schools preparing the boys to get into Rugby responded to Temple's changes by starting their own, similar, exams.[100] People's expectations were being pre-structured as they became increasingly practised at taking – or giving – examinations. The river of testing not only cut its own bed: over time, it cut that bed deeper.

Exam Systems Spread Overseas

The year 1858 thus marks the end of the first great expansion of different exam systems across the British Isles. After 1858 these systems extended in three new directions: *downward* to schools for working-class children and adults; *outward* to places throughout the British Empire; and *otherward* to members of groups not hitherto seen as commensurable with males – female candidates in particular.

In 1858 the University of London extended, and Cambridge University created, their own exam systems. The University of London Matriculation exam was often written as a "school-leaving" (i.e., graduation) test for senior students, who went to London to write it; after several requests, in June of that year the Matriculation exam was written, simultaneously, at five different schools in places like Manchester. The senate issued extremely detailed instructions to its examiners, which specified the physical distance between candidates, the order of places in which they were to sit, the timing of the exam, and the recording of registration information. Examiners were also given the requisite supplies, and the importance of how these papers were to be handled was high, given the frequency of the word "sealed" in these instructions.[101]

In December 1858 Cambridge began its own Local Examinations for candidates under sixteen and under eighteen; unlike Oxford's, the Cambridge Locals were seen as preparatory to university.[102] They quickly become more popular than Oxford's because from their inception the Cambridge Local exams had a larger pool of candidates: not only boys from old grammar schools, but also from national schools.[103] Many were reluctant to write the Oxford exams because they were seen as "more exacting."[104] Moreover, Oxford organizers were less aggressive at starting up new exam centres, and they insisted that one of their main examiners should attend at each examination centre during writing. By 1880 Oxford Locals were being written by only a third of the number of candidates who took the Cambridge ones.[105]

The Cambridge Local Examinations also came to be the second test written overseas. On 17 December 1862 the Cambridge Local Examination Syndicate was asked to hold the exam in Trinidad; it accepted, and ten candidates from Queen's Collegiate School[106] wrote in Port of Spain in December 1863.[107] While it is unclear who made the request to the Examination Syndicate, the likely person was Horace Deighton, headmaster of Queen's Collegiate School and twenty-first wrangler of 1854, Queens' College, Cambridge.[108] From 1866 on, students at Queen's Collegiate took Cambridge Local exams to determine who earned exhibitions (scholarships) to pay for their tuition.[109] Since that school was secular, modelled upon Queen's Colleges in Ireland, its nickname was the "Godless" or "English College." Its rival school, St Mary's College (the "French College"), was Catholic and francophone, with a higher proportion of Black and Creole students. Yet both schools taught Classics and mathematics, as well as topics such as history, chemistry, and modern languages.[110] So St Mary's College students began to write the Cambridge exams as well. Because each school represented a larger faction in Trinidadian society, there began a rivalry between Queen's and St Mary's over whose students could place the highest on the Cambridge Local exams. The teaching at both schools accordingly shifted to emphasize the topics asked on the Local exams, particularly once scholarships to Cambridge were awarded to the two Trinidadian boys who scored highest on the senior exams. There were sensible criticisms that these exams did not focus on commerce or agriculture, and that what resulted was a "crazy-quilt" curriculum that was too focused on books and unconnected with much that was relevant to most Trinidadian lives. But for many Trinidadians what mattered was not only that the Cambridge

Local Examination set out entrance requirements to British universities – it had also become an entrenched standard allowing comparisons to be made between schools both local and across the British Empire.[111]

Far away in the Indian Ocean, in July 1864 the Royal College of Mauritius requested that its "highest candidates" write the London Matriculation Examination citing the precedent of its being written at provincial universities in the UK. The Royal College was Mauritius's only secondary school. The London Matriculation seems to have been requested by a new guard of émigré teachers, notably Walter Besant, eighteenth wrangler, Christ's College, Cambridge, 1859, and Frederick Guthrie, BA 1852, University of London. They wished to show that the college had changed its course after a staff revolt against a detested headmaster. The Matriculation would determine who was eligible for scholarships to English universities, but more importantly it would serve as a "constant and reliable test" independent of the college teaching staff.[112] The university senate agreed, as long as the exams were taken in conditions as close as possible to how they were written in England. They were run by a Mauritius official not connected to the Royal College; he handled the sealed questions when they arrived from London, oversaw the exam, and then sent them back in sealed envelopes to London for marking. The performances were dreadful. In 1865 only six of 341 pupils passed; the next year it was three of 247; in 1867, three of 164.[113] One reason for this was a widespread cholera epidemic, which made the exam conditions decidedly non-uniform. Nonetheless the Mauritius exam set a precedent: in 1866 London's senate declared that any Matriculation exam written under the same conditions as the Royal College's would be recognized by the University of London: "By this system, the most perfect uniformity is ensured," explained London's Registrar, W.B. Carpenter, in response to a request from the Dean of Cape Town that the Matriculation exam be written there.[114]

The two exam systems spread ever further outward. By the late 1870s London Matriculation exams were used to award prizes and a new award, the Gilchrist Scholarship, for colonials wishing to attend the University of Edinburgh or University College London. By the mid-1880s the Matriculation was being written in cities like Bombay (Mumbai), Hobart, and Montreal.[115] Meanwhile the Cambridge Local exam system had also continued to spread to places as far-flung as Auckland, Colombo, Georgetown in Guiana, Madras (Chennai) – and in Mauritius too.[116] There was so much demand that the Cambridge Syndicate, wary of other universities' jurisdictions, turned down requests to run it where other universities were dominant, such as Toronto, or in places outside the empire, such as Albany, New York.[117]

By the 1890s the extension of London's and Cambridge's examination networks – and the creation of rival systems by colonial universities, for instance in the University of Toronto's own Local exams for women[118] – had made exam writing a taken-for-granted fact among many in the British Empire. And the belief that examinations were a tool of British hegemony was nicely summed up in a poem dedicated to the then-head of the Cambridge Examination Syndicate, John Neville Keynes, father of John Maynard:

> Though Roman legions ruled the world,
> Though Britain's thunderbolts are hurled
> At Monarchs in Ashanti Plains;
> The Locals Syndicate preside
> O'er realms more gloriously wide,
> Broad as the sky are their domains
> Black babes or yellow, brown or white,
> Cram manuals from morn to night
> No hue from culture now refrains;
> The infant startles from his cot,
> His bottle and his bed forgot,
> To moan aloud the name of K[eynes].[119]

Whether in the British Isles or across the empire, examinations had come to shape people's expectations and imaginations about what schooling – and education more generally – ought to be.

"An Adopted Chinese Culture"

Exam scores rose throughout the 1860s. George Dasent, an examiner in English language and literature for the Council of Military Education and the Indian Civil Service, had failed eighty of the one hundred candidates for the Woolwich entrance exam in 1855. By 1866, he failed only seven of about a hundred candidates. Dasent believed that teaching had improved ("teachers have taught themselves"), and that candidates now knew far more about English language and literature. But he suggested that exam results had also greatly improved because candidates had polished their own exam responses into formats more likely to satisfy an examiner. The first exam papers had the "most ridiculous scrawls upon them" and candidates' syntax was jumbled – but over eleven years this had greatly changed as candidates learned how to best demonstrate their knowledge on examinations.[120]

The greater emphasis on examinations started to redefine schooling and education, complained some. Such grumbling began as early as 1837, when William Whewell, Master of Trinity College, noted that students were coming to define a Cambridge education only in terms of examinability.[121] In 1860 Thomas Love Peacock's *Gryll Grange* noted that exams would have excluded Marlborough from the army or Nelson from the navy.[122] Old Indian Civil Service hands worried that the new emphasis on exams led to far more bookish civil servants, better known as examination or competition "Wallahs."[123]

Meanwhile, more specific concerns about the overuse of exams appeared. In 1859 one teacher pointed out that the existence of five separate exam systems, each with different requirements, confused teachers and students about how and what to teach and learn.[124] There were also other concerns that exams rewarded memorization without comprehension, or "cramming." Such fears grew in 1862, with the adoption of "payment by results" – tying funding to examination scores – for government-funded national schools. In 1867 the Rugby science teacher James Wilson noted that because the study of science was so often structured by exams, topics like geology and chemistry were "frightfully crammable."[125] Observations like these were especially dangerous to people like T.H. Huxley who promised that the study of science would sharpen students' inductive abilities.

As discontent with exams grew, there came claims that Britain was becoming "Chinese." In the 1850s there had been neutral comparisons of British exams with China's: in 1853 Earl Granville – later, Lowe's political master at the Committee of the Privy Council on Education – noted the success of the Chinese system of competitive examinations. Did the notion of competitive examinations come from China? Some believed it did: in 1857 a writer for the *Westminster Review* noted that French revolutionaries, by emphasizing talent, introduced "the long-established practice of China"; and with its new civil service exams, the English were also "taking a leaf out of Chinese books."[126] The sense, here, was that the exams were providing some upward mobility to talented youth, as with the Qing exams. However, examinations also served another purpose in the Qing empire: to produce a large number of talented people who thought in a common way, thereby unifying people from a diverse set of cultures across China.[127]

It is this second sense, that examinations caused indoctrination, which began to create discomfort in the 1860s. It was accompanied by a larger European shift of condescension towards China.[128] Lowe, as the politician most directly responsible for the policy of payment by results, and as MP for the University of London, became known as the "British Confucius."

In 1862, his foe Matthew Arnold decried payment by results by comparing it to the Chinese system. Lyon Playfair, who had also done much to create payment by results, had radically changed his beliefs by the 1870s. He declared that the University of London was inferior to other universities because it resembled the examining boards of China, whose system had produced (and here he paraphrased Ernest Renan) a "general and incurable senility."[129] On the subject of university reform, liberalism was accused of standing "chiefly mute, fearing lest its Chinese idol, competitive examination, should be overthrown, and his joss-house burnt in the struggle."[130] A.H. Sayce's 1878 warning of an "adopted Chinese culture" in Britain has already been noted.[131] John Lubbock, as he distributed the March 1879 prizes of the College of Preceptors and praised the science teacher Robert Goffin, mused that just as exams monopolized public attention in China, they were beginning to do so in England. By 1884 a review of a textbook in experimental chemistry praised the book's avoidance of the word "examination" and hoped that it prophesied a future when there had passed an over-reliance on examinations, which at that point "earns for us the title of the 'Chinese of Europe.'"[132]

By the 1880s, competitive examinations were a subject of some comedy: Gilbert and Sullivan's *Iolanthe* features an MP's bill to have the peerage selected by competitive examination.[133] More seriously, the politician Auberon Herbert, who in 1880 had complained that exams were turning schooling into a "great copy-book,"[134] was by 1888 organizing a petition against "over-examination" warning of the "dangerous mental pressures" they caused. The petition received thousands of signatures from various eminent residents of the British Isles.[135] By 1890 Frederick Pollock was more thoughtful about the topic – the mania for exams had been part of the "useful knowledge" movement, which assumed that teaching would emerge spontaneously to meet demand for exams. This had not occurred, however, and examining by itself did not constitute a "liberal education."[136]

By this time the belief that examinations could both test and instil moral virtues had come to be ridiculed. One 1898 history of schooling in England concluded that an over-reliance on examinations had actually become *destructive* of morality, particularly when an exam was tied to earning grants. Such "money-grubbing" intentions forced children into the role of "money-making machines out of which the last penny must be squeezed."[137] Just how exams became devices with which to raise funds, and children into machines, is the topic of the next chapter.

2 Monetizing Marks: The Political Economy of Examinations

Political economists, pure and simple, conceive mankind as animated by the one desire of buying in the cheapest and selling in the dearest market; they find that it conduces to clearness of thought to eliminate for the time all other motives. In the same way … I should like, for the nonce, to think of the boy simply as a creature desirous of marks, and his teacher as the just distributor of them.

– H.W. Eve, "On Marking," 14

The "good angel of humane and rational education, Sir J. Kay-Shuttleworth, was succeeded by the evil genius of beggarly elements and payment by results, Mr. Robert Lowe." This was the 1898 view of historian of education Henry Holman, who considered payment by results to be one of the most important single developments in the past half century of schooling in England. And it was a dreadful one, for it raised "the elements of machinery and the mechanical … to their highest powers."[1] Views of Lowe as a demonic figure persist even to this day, given his reputation as the architect of funding a school according to its students' exam results. Americans now know such funding schemes under designations such as "No Child Left Behind," but these systems were first pioneered in England.

The recurring image of exams as "mechanical" shall be visited in a later chapter; the topic of this one is how there emerged a way of deploying exams that not only purported to measure student attainment, but that also became tied to an invisible hand of funding education. The emergence of payment by results shows how public policy was informed by deductive models of political economy. Lawrence Goldman has noted how such models, specifically Ricardian economics, gave decision-makers the ability to see developments through a "deductive system of beguiling clarity."[2] School reform and the 1850s push for "free trade in education" used the same language as that of the campaign leading to the 1846 repeal of the Corn Laws.

The emergence of payment by results also shows how abstractions can simplify a messy policy realm understood only by experts. By making a complicated system "legible,"[3] abstractions make that system accountable to outside authorities, even if those authorities themselves do not fully understand it. Such legibility is important if there is to be democratic accountability, although there is always the danger of politicians and civil servants coming to focus only on the abstraction, forgetting about the underlying complex territory that this map represents. With payment by results, the "evil" Lowe sought to make a schooling system more accountable to the non-expert MPs who voted to give it money. And Lowe also wished to ensure that schools receiving public funds focused on all their students, not just the visibly bright ones.

Lowe, however, was not the actual creator of payment by results: that honour goes to the Department of Science and Art. The DSA is best known to historians for its science instruction facilities in South Kensington, especially the teaching laboratories created by T.H. Huxley, Edward Frankland, and Frederick Guthrie. Such institution-building at "South Kensington" has been discussed by Sophie Forgan and Graeme Gooday. And it did indeed revolutionize science teaching across the British Empire and even in the United States, especially Huxley's emphasis on laboratory work in learning science, for thousands did study at these teaching laboratories.[4]

Yet if one is to assess influence by numbers alone, the most important work done by the DSA is shown by its testing over two million candidates between 1861 and 1900.[5] The growth in the number of candidates taking DSA exams was extraordinary, starting with 650 candidates in May 1861 and peaking at 114,632 in May 1898.[6] On average each candidate wrote two exams, but for each person who wrote a DSA exam, there were two or three others who did not and yet still followed the DSA syllabus.[7] It is thus a matter of some understatement that over their thirty-nine-year run the DSA exams were an important way to get people to learn about science. As historian David Layton observes, the DSA exams really did mean "science for the people."[8] They shaped mass perceptions of science and thus merit the attention of anyone interested in how science was "popularized." And the DSA's pioneering use of paying teachers for their students' exam results may have contributed to the notion that science is "useful" – in addition to revealing truths about the world, it paid better than other subjects. Even to get into the South Kensington teaching laboratories, students had to do well on the relevant exam and on its second level of difficulty, its "advanced" paper.

Hence these exams served not only to test thousands of people a year of their knowledge of science, but also to identify those who showed promise in it. Huxley's exams can be said to have "discovered" Margaret McConnish, his first female laboratory demonstrator, and H.G. Wells. The skills of Oliver Lodge and John Perry were identified by John Tyndall's exams.[9] And doubtless the DSA exams interested many others in science who otherwise would have been unaware of it.

Yet apart from the work by Harry Butterworth, William Brock, and more recently Ruth Barton, little has been written specifically about the DSA exams by historians of science. Scholars may have been deterred by the complexity of different schooling systems in the British Isles; moreover, those who have studied the popularization of science have tended to overlook schooling, to repeat a point made by Josep Simon.[10] Yet understanding the DSA exams is crucial for understanding the standardization of exams and why so many educationists were inspired by deductive models of political economy. Thinking about education as governed by the forces of supply and demand was what inspired the first widespread system to give money for good exam marks – the DSA's policy of payment on results.

Montagues and Capulets

The previous chapter noted the disunity of different schooling systems in the British Isles. Administration systems were equally complicated. By 1856, in England and Wales alone, there were *two* government branches that tried to regulate schooling by inspecting and distributing government monies: the Education Department and the Department of Science and Art. They were both governed by the Committee of the Privy Council on Education.[11]

The Education Department, centred in Whitehall, focused on national schools. It had emerged out of James Kay-Shuttleworth's 1839 proposal to train and certify teachers, and in 1846 it also took on the administration of apprentice pupil-teachers. It employed Her Majesty's Inspectors (HMI), who included such eminent figures as Matthew Arnold and Frederick Temple. Above, we saw how by the early 1850s Balliol College graduates were strongly represented among the ranks of HMIs, partly due to the influence of Balliol's Benjamin Jowett, but also of Ralph Lingen, a Balliol graduate who succeeded Kay-Shuttleworth as the Education Department's permanent Secretary.

The DSA, located in South Kensington, emerged after the Great Exhibition of 1851. It was initially part of the Board of Trade, since its mandate was to improve the manufactures of the British Isles by helping the working classes learn more about science, art, and design. The

1856 move of the DSA out of Trade and into the Committee of the Privy Council on Education led to an uneasy coexistence with the Education Department. They did not merge: rather, the Education Department was given responsibility for "general education" and the DSA for "industrial education." Politically they were controlled by the Lord President and Vice President of Council, but each group remained independent; only in 1895 would they be unified.[12] The two groups were thus often depicted as being in opposition, with one HMI claiming their members never spoke with one another; instead, they were like "Montague and Capulet retainers [who] never met; only they bit their thumbs when the other house was mentioned."[13] Although entertaining, the description is not quite accurate. New administrative schemes and personnel were shared between the two departments, especially through the secretaries, Vice President, and Lord President. There certainly were differences between the two groups – rather than Balliol men, or Oxbridge graduates more generally, the DSA tended to employ Royal Engineers and London men of science. Not only is this a case of the "two cultures" – literature versus science – it also shows the Oxbridge-Metropolitan divide identified by such scholars as Nicolaas Rupke.[14]

In 1861, new regulations stipulated that a school could receive money from either the Education Department or the DSA, but not both.[15] Lingen had moved to prevent any DSA subsidy for science and art classes in national schools. As a result the DSA focused its efforts on supporting evening classes and non-government-funded schools. Yet Lingen's restriction was frequently ignored by individual teachers who did end up getting paid by both groups. Moreover, although DSA rules specified that the classes it funded were intended only for workers and children of the labouring poor, such regulations do not seem to have been enforced. The money that teachers stood to gain from getting their students to pass DSA exams was incentive enough for them to ignore such restrictions.

Money also seems to have been the initial attraction for the eminent men of science who drew up the DSA exams at the end of the 1850s; the DSA deliberately linked the reputation of its exams with the reputation of its examiners. Through the 1860s, as the DSA exams grew in size and importance, London men of science such as T.H. Huxley and John Tyndall realized that the exams offered benefits beyond spreading greater knowledge about science throughout the population: the exams helped gain cultural authority for the claims made about science by those men. Many of the DSA examiners also happened to be members of the infamous "X Club," a dinner society founded in 1864 that met monthly to plan control of science in the metropolis. Members of the X Club who were not themselves examiners were linked with examinations in other indirect ways: for instance, X Club member William Spottiswoode, a mathematician and printer, had been tutored by Frederick Temple at Balliol, where he became friends with Benjamin Jowett.[16]

Yet only one of the three most important DSA figures was truly a man of science, and he was not even an X Club member. This was Lyon Playfair: more about him below. But first we shall look at the two other architects of the system: Henry Cole, known as "Old King Cole" for his mastery of Victorian bureaucracy, and the Royal Engineer John Donnelly, who did much to work out the details of the DSA exam system. All three of them took the emerging system of mass examination and tied exam results to funding. These three men discovered how to monetize marks.

"Teaching Is a Drudgery"

Henry Cole was an innovative and sharp-elbowed administrator, skilled at bureaucratic infighting and pamphleteering. With John Stuart Mill and Edwin Chadwick, Cole was a philosophical radical, a civil servant and journalist who admired the utilitarian principles of Jeremy Bentham and James Mill. Cole, along with Chadwick, has been depicted as part of the "revolution in government."[17] Cole first made his name in 1835 when as assistant keeper in the Record Office he campaigned for accurate records of witness testimony at parliamentary commissions. He charged that exchanges that embarrassed someone tended to disappear or be changed when recorded on paper; this often happened after the notes were sent to that witness for clarification.[18] Such a preference for written records – a favourite cause of philosophical radicals – recurs throughout Cole's career and shaped the DSA exams.

The standardization of DSA exams was informed by another cause important to Cole and other philosophical radicals: reducing waste by "securing uniformity." Cole's interest in efficiency was shown by his work on the Penny Post and railway gauges. From the late 1830s onward he championed Rowland Hill's Penny Post system, which emerged from Hill's claim that the greatest cost of postage came not from distance travelled but from all the different duties performed by each clerk in sorting, checking postage, and balancing accounts.[19] Because a clerk had to constantly switch tasks, the system was prone to errors and fraud.[20] The famous prepaid stamps of the Penny Post were simply the most visible part of the postal reforms, which also relieved overburdened clerks of having to also balance accounts. The simplification of their duties made clerks more efficient. Cole believed standardization was "key to Hill's great discovery," and it is telling that when Cole agitated for Hill's proposal in the 1830s he always added the word "uniform" to the words "Penny post."[21] From the mid-1840s on, Cole pamphleteered for uniform railway systems during the "Battle of the Gauges." In 1846, there were two different railway gauges across England and Wales – one 4′8″ wide, the other 7′. The 4′8″ "narrow" gauge was the traditional one, used by a majority of the railway companies expanding across England and Wales; the upstart 7′ "broad" gauge was used in the southwest of England.[22] A railway engine and car built for one gauge could not travel on the other. Where the two rival gauges met, at Gloucester, all passengers and cargo had to be transferred, incurring inconvenience and additional costs. Cole's pamphlets even estimated the expense per ton this gap caused.[23]

By June 1849 Cole was stumping for the forthcoming Great Exhibition; it was his suggestion to open the event up to exhibitors from other countries, which signified a victory for free trade. His exhibition work made Cole a leading member of the Society of Arts; in January 1852 he was offered the post of Secretary of what became the Department of Practical Art.[24] Cole's task was to manage Schools of Design that taught scientific principles to British artisans. He soon came to believe that many of these schools' teachers were incompetent at teaching and even drawing.[25] Moreover, graduates of the Schools of Design had trouble finding work: for instance, although four porcelain painters trained at the schools had won awards for their work, only one was ever hired by a manufacturer, and at below-average wages.[26] To explain why, Cole turned to deductive political economy: the graduating porcelain painters were never hired because there was "no adequate market" for their skills.[27]

The use of such deductive models Cole shared with others, like Lowe, who had called for more competition between Oxford's colleges to make the university more efficient.[28] Cole saw relationships between student, teacher, and knowledge in terms of supply and demand, competition, and efficiency. Graduates of the Schools of Design were unable to get jobs because they lacked a "market" for their services; potential employers were one source of "demand" for those graduates. By implication, then, teachers and schools could be characterized as "supply." The language of political economy also gave Cole a way to explain a problem faced by the DSA during the 1850s: it had too many teachers and not enough students, particularly in science. As early as 1853, the Department had given piecemeal aid to schools teaching science, but these schools routinely failed for lack of students.

This belief in a lack of demand for science teaching was also borne out by evidence from outside the DSA: when in 1859 a chemistry teacher advertised his services in the newspaper to London-area middle-class schools, he received no responses.[29]

Deductive economic models tend to smuggle in beliefs about human nature,[30] such as the assumption that everyone wishes to work less. When it came to teaching, Adam Smith argued that every teacher sought to make the most amount of money while doing as little as possible. This point was repeated by John Stuart Mill: if a teacher's wages remained unchanged, then it was in their interest to "have as few pupils as possible, and to take the least possible trouble with their instruction."[31] In this view, it was economically *rational* for a teacher to do as little as possible. Donnelly (who himself had never taught) explained why: teaching was a "great drudgery."[32] For his part, Lowe (who had) claimed that "teaching is a trade, and not a very highly intellectual trade. There is no occupation more likely to degenerate into lifeless routine and meaningless repetition. To be perpetually saying the same things to different people, to explain the same difficulties, to use the same illustrations, requires some extraordinary stimulus from without to prevent it from degenerating into something as useless to the learner as it is intolerable to the teacher."[33] The "extraordinary stimulus from without" was the exam.

Since it was assumed that it was in the interest of teachers to work as little as possible yet in the interests of society for students to learn as much as possible, a device that aligned these two opposed interests was necessary. This device was the twinning of examination results with payments for teachers. Turning to Mill again: "The true principle for the remuneration of schoolmasters of all classes and grades, wherever it is possible to apply it, is that of payment for results. The results of their teaching can, in general, only be tested by examinations, conducted by independent public examiners."[34] Mill's insistence upon *independent public examiners* brings us to a further central Victorian assumption about how examinations ought to work: they needed to be kept separate from teaching. Teachers frequently complained that examiners did not know anything about teaching, which meant that their exams were too often set at too abstract or too difficult a standard. Yet this was entirely the point when seen through the lens of political economy and beliefs about the proper alignment of interests. Seen through another lens – that of the law – judging was separated from advocacy, and examining was a "judicial act."[35] This separation of teaching from examining was reinforced by the fractious schooling environment in the British Isles, and the belief that schools were better when they were set up by neighbourhood community members rather than government departments. With no curriculum set by a central ministry, external examinations would set standards about what an educated person ought to know.

Reinforcing the distinction between teaching and examining was social status. Teaching was already seen as a lower-status occupation. Examiners had a higher status than teachers not only epistemologically – knowing more about a given subject – but also socially. While teachers might aspire to gentility, and "resent" their lower social status (a point made frequently in various reports penned by Oxbridge-trained HMIs), their lower status was always assumed. Although both the Education Department and the DSA were separate institutions, both groups' members agreed upon a central point: teaching was a trade.

Legibility and Elasticity

A challenge for Cole was inconsistent funding for education. In some years, Parliament would vote for large sums to be given for schooling. In other years, after calls were made for financial retrenchment, it overcorrected by severely cutting back those sums. Cole saw examination results as one way to make parliamentary funding more consistent. In modern terminology, exams and their results made the complicated schooling environment more "legible." Politically, there was an additional benefit to the use of examinations: it was easy to raise standards by making exams more difficult, and such changes could be clearly depicted as an improvement by decision-makers to members of the public.

Between 1852 and 1854 Cole, Harry Chester, and James Booth sat on the Council of the Society of Arts, learning how to run mass exams for artisans at Mechanics' Institutes. On the council it is likely that Chester explained both the Education Department's certification examinations for pupil-teachers and teachers and the department's subsidy of teachers' incomes by the level of certificate they gained. In 1853 Cole declared that prospective art teachers for the Schools of Design would no longer be hired on the strength of their testimonials. Instead they were to demonstrate their competence by writing a preliminary exam that involved answering written questions, solving problems in geometry and perspective, drawing and painting certain objects such as shaded solid models and foliage, and teaching a class. Passing the first exam meant not only entry to the schools but also the chance to gain further certificates that paid more money. By March 1854 Cole had copied the Education Department's system of supplementing art teachers' incomes according to their certificate level – up to £50 a year.[36]

But now Cole added a new feature: pupil-teachers who passed drawing exams would also earn money for the teachers who had taught them. What drove Cole to this decision? Such bounties fit nicely with Cole's belief that economic efficiency was gained through piecework: had he the power to do so, he would have had all civil servants paid this way, since Cole believed that "payment by salaries independent of the results of work" contradicted economic laws.[37] His friend Chadwick had used piecework during his time at the Poor Law Commission to cut the number of clerks from seventy to thirty-five; they could copy as much as they liked and ended up increasing their incomes from £60 and £70 to up to £100 a year.[38] The financial interests of the workers and Chadwick's own desire to make his office more efficient were aligned. A love of piecework was satirized in a *Punch* article claiming the DSA was experimenting with paying a "fixed price per job": naming a fossil earned a worker one shilling; half an hour's conversation with a visitor paid two shillings and six pence.[39] At any rate, by May 1856 three shillings were paid to any teacher whose student won a prize on an art exam, and in 1857–8 the DSA made 5,259 of these payments.[40]

To introduce our second important DSA figure, Lyon Playfair, we need to jump back to March 1853, when the Department of Practical Art became the Department of Science and Art. Cole became joint Secretary with Playfair: Cole was in charge of art, and Playfair, of science. Indeed, for a clearer picture we need to go back even further: like Cole, Playfair had been in close contact with philosophical radicals as a youth. In the late 1830s, when he studied medicine and chemistry at University College London, he came into contact – possibly via fellow Scot Neil Arnott – with Rowland Hill, John Stuart Mill, and Edwin Chadwick. These "friends in council" dined at one another's houses and discussed economic and postal reforms.[41] Playfair then went to Germany to gain a PhD with the world's most famous chemist, Justus von Liebig. When he returned to the UK, Playfair became Liebig's British representative, which gained him the notice of influential figures such as Prince Albert and Prime Minister Robert Peel; Playfair eventually became a special commissioner for the Great Exhibition. Playfair was perhaps the first public figure to warn about the economic consequences of neglecting what we now call "STEM" education: in 1852 he claimed that British industry, due to its neglect of scientific principles, was in danger of being surpassed by industries on the Continent.[42] Such economic alarms were one reason why the teaching of science and art was placed under the governance of the Committee of the Privy Council on Education.

The DSA used profits from the Great Exhibition, and a government loan, to buy land in South Kensington to establish museums and offices.[43] It was also in charge of London's Museum of Practical Geology and the Government School of Mines, as well as Dublin's Irish Industrial Museum. Thus it not only paid the salaries of professors like T.H. Huxley – the DSA also paid for public lectures at the Irish Industrial Museum.[44] Under pressure to cut costs, Playfair looked at the value of the Irish public lectures. Already in favour of requiring exams to gain government employment,[45] he asked that paper exams be administered after a series of lectures at the museum. He anticipated that the scores would be low, and thereby justify cutting the Irish lectures. These exams were administered in 1855–6. While the exams were not mandatory for attendees, certificates and prizes were offered to attract candidates. The results surprised Playfair: the lectures *had* been successful at teaching attendees scientific knowledge, or more precisely, the prospect of an exam had "forced the pupils to read & study & the answering has been surprisingly good," he confided to a correspondent. Three female attendees had performed well, with one gaining the first prizes in botany and zoology.[46]

Playfair decided that what mattered was not the specific *process* by which one learned science; rather, it was the *results* of that learning. He concluded that it was best if the DSA did not dictate pedagogical methods – it ought only to assess whether correct learning had occurred. By focusing on rewarding results, the DSA would give "private enterprise" the chance to experiment with the best ways to teach science. In short, Playfair stated that the DSA no longer cared if science was taught in a school or in a "garret, by books, oral demonstration, or Experiment."[47] The Irish museum experiment showed that an exam could not only be used as a measurement device and as a tool for reform; its results could also be deployed as evidence by a bureaucratic agency under external pressure to defend or change an existing policy. In this defensive quality, exams seemed to share something with how statistics could be used – possibly because exams were generating those statistics. The same applies today.[48]

By 1856 Cole and Playfair had enlisted new supporters of their exams. A group of young and ambitious men of science were discovering the value of becoming examiners. Since 1854 the botanist Joseph Dalton Hooker had been examining assistant surgeons for entry into the Indian army, and in 1855 he became an examiner for the botany medal of the Apothecaries Company. In April 1856 he suggested to his friend T.H. Huxley that they could reform natural history teaching if they had "sufficient command over the public, as examiners in London, and as confidential advisers of examiners and professors elsewhere, to ensure the cordial reception of such a system." When combined with Hooker's later declaration to Huxley that they would form a "committee of public safety" to govern science,[49] there seems to be conspiratorial sinister intent, especially in the light of the formation of the X Club in 1864. But there is another, more prosaic, explanation for their embrace of examining: the jobs often paid quite well, supplementing their meagre salaries. Huxley succeeded W.B. Carpenter as the University of London's anatomy and physiology examiner when Carpenter became the university's registrar, while Huxley's friend John Tyndall became an examiner for the Council on Military Education.

In June 1858 Playfair took the chair in chemistry at the University of Edinburgh, leaving Cole as the sole Secretary of the DSA. The administration of science for the department he gave to a twenty-four-year-old Royal Engineer and decorated Crimea veteran, Lieutenant John Donnelly. As with the wranglers and double firsts who equated education with examination success, Donnelly was similarly well disposed to exams, having entered Woolwich in 1849 as the top-performing candidate on its entrance exam.[50] Meanwhile, Cole continued to tinker with the examinations of the Society of Arts, also praising the Oxford Locals as a model that allowed exams to be written wherever a board had requested them.[51] Governments came and went, with political masters of widely different levels of interest; some seemed "disposed to leave details to us,"[52] but by the spring of 1859 a new Lord President of the Committee of the Privy Council on Education, the Marquis of Salisbury, was concerned that so little science was being taught by the DSA that he might remove the "science" portion of the department's name. To expand the DSA's offerings in science, Cole saw exams as a solution, and between April and July 1859 he plotted with Donnelly, studied the workings of the new Oxford Local exam system, and met with scientific figures such as Michael Faraday, August Hofmann, Tyndall, and Huxley; the latter three all taught at the DSA's School of Mines. Another change of government meant that Salisbury was replaced by Earl Granville as Lord President and Robert Lowe as Vice President.[53]

Cole's request that DSA students simply take the science exams of the Society of Arts was denied, and so he decided to establish the DSA's own exam system. Teachers would gain money when their students passed these exams. By November 1859 the DSA had hired specialist examiners from its School of Mines such as Tyndall (physics), Hofmann (chemistry), Andrew Ramsay (geology), Huxley (natural history), and Edwin Lankester (botany). Each examiner drew up a syllabus of examinable topics, then questions for a teachers' exam. This first teachers' exam was written in June 1860 by roughly 500 candidates; teachers who passed were entitled to a payment when one of their students passed a different, general, exam. By 1865, the DSA decided that having separate teachers' exams was too cumbersome and simplified the process. Anyone passing an exam on a topic could claim a payment from the DSA if they taught another candidate who subsequently passed that same exam. The candidate must have registered in advance as that person's student and attest to having received at least twenty-five lessons on the topic. Thus if someone passed the elementary paper of T.H. Huxley's animal physiology exam, the certificate gave them the right to claim DSA money if one of their own registered students later passed that same elementary paper.[54]

Bureaucratic tussles did cause some additional complexities: when Lingen prohibited anyone in his Education Department from helping with DSA exams, Cole turned to the Royal Engineers for help. Not only was Donnelly an engineer; Cole had worked with the Royal Engineers since 1849, in preparation for the Great Exhibition, and favoured them (his own son became one).[55] At the DSA, the Royal Engineers established their own parallel inspectorate to the Education Department's HMIs, inspecting both exams and the schools in which science was taught. Donnelly's assistant, Captain William Abney, thought that every DSA teacher should teach with the expectation that an inspector might visit at any moment.[56]

The 1860s and 1870s saw the rapid expansion of the DSA exams. In 1864 T.H. Huxley told Hooker that exams were "the most important engine yet invented for forcing Science into ordinary education"; by 1868 Huxley argued that the system was "one of the greatest steps ever made in this country towards spreading a knowledge of science among the people" because exams spread science from "below upwards" rather than "from above downward."[57] By 1869 he noted that examiners could cause an "almost unlimited improvement" in learning science as the system was refined.[58] Huxley became good friends with Donnelly,[59] and was so strong a supporter of the DSA that in 1887 he wrote a letter of support to *The Times* that even the highly sympathetic members of *Nature* thought a little too one-sided.[60]

Perhaps the most important legacy of the DSA examination system was its influence overseas. Since the 1870s the Government of India wished to extend scientific instruction to its population as well. In 1885 H.B. Grigg, director of public instruction for the Government of Madras (Chennai), suggested they follow the DSA model. Using economic language, Grigg argued that its success showed that payment-by-exam results could create a large enough market to employ competent science teachers; moreover, it had helped England, which had fallen behind continental science schooling, to catch up. In addition to syllabi published in advance by public examiners, candidates for these exams could use as one standard the Matriculation exam of the University of Madras.[61] This scheme was approved by the Government of Madras, with each student getting eight rupees for a second-class result on a preliminary higher exam, and twelve for a first class. "Fifty per cent. [*sic*] extra on the above rates will be given for girls and Muhammadan boys," the government declared,[62] conveying its desire to encourage previously overlooked groups to take science lessons. From Madras the system spread to other parts of India.

Invisible Hands

The guiding principle of payment on results was what administrators called "elasticity." Cole believed that with this he had solved the problem of unstable parliamentary funding. The prospect of financial reward would induce many to go into science teaching, with the most efficient teachers making the most money. Once a large enough group of science teachers had emerged – retaining enough students for those teachers to continue making a living – the DSA could cut back on its funding. Donnelly also hoped that over time, DSA standards could be raised simply by making the exams more difficult. Because the exams were set not by teachers but by experts in each subject, standards could not be lowered to suit the teachers. There was no conflict of interest.[63]

Deductive models of political economy, then, helped shape the DSA's exam system. Such a system then came to be adopted by the DSA's counterparts at the Education Department in the Revised Code of 1862 – a system better known as "payment *by* results." The links in common were Cole, Donnelly, and Lowe. As Vice President of the Committee of the Privy Council on Education, Lowe was charged with reducing the budget of the Education Department: while it had begun in 1834 with an initial outlay of £20,000, by 1860 its budget had grown to £886,920.[64] Chancellor Gladstone believed his ally Lowe was ideal for this task, with a talent "for unravelling and penetrating a system of waste and fraud."[65]

Lowe was struck by the complexity of the administrative system of the Education Department. He saw an office beset by far too much paper and overwork – a condition blamed for James Kay-Shuttleworth's 1849 breakdown.[66] As with the pre–Penny Post Office, the Education Department's systems were seen as too complicated. One problem was the volume of correspondence: in 1859 the Department was in direct contact with 15,224 pupil-teachers regarding their payments, and as a result they were frequently paid up to a year late. By 1862, the

Department was in contact with a total of 38,331 people, and it was necessary to "handle" each of them. To illustrate, Lowe gave one example of a dispute over a £15 payment to a single pupil-teacher: it required twelve letters, an hour-long visit to the Department by a school manager, then a thirteenth letter restating everything again.[67]

Another problem was how HMIs wrote their reports, which followed no uniform format. The reports could run to hundreds of pages, were often didactic, and a few even proposed new legislation. Each year these reports grew slightly longer. Even before Lowe's time there had been struggles between the central office at Whitehall and the HMIs over how the reports ought to be more uniform. Adding to the complexity were the different interpretations of standards among HMIs: they would classify a school as "excellent," "good," "fair," or "bad" but often held wildly different views about what excellence or badness actually meant.[68] In modern parlance, administrators of the Education Department had to handle an enormous amount of raw, and untabulated, information about schools' performance. This information could not be simplified – that is, abstracted – in a meaningful way. For decision-makers such as Lowe, this lack of legibility meant that the department could not be accountable to Parliament.

Lowe faced additional pressures when the Newcastle Commission issued its report in 1861, after three years of investigating the condition of government-funded national schools. The report's most publicized finding was that much of the public money spent on education was wasted. The commission blamed how national school students were assessed: school inspections hadn't kept up with the growth in the number of numbers. At first, students were assessed *only* through inspection. When inspections started in the late 1830s, the scale was small, with HMIs told to be "ameliorative" – inspections were supposed to ascertain the intellectual, moral, and religious progress of the students and suggest ways for each school to improve.[69] Yet this method hadn't changed despite the explosion in the number of schools. As a result, an HMI usually only had time to inspect each national school in its entirety: checking each class, scrutinizing attendance registers, examining the condition of the building and grounds, observing pupil-teachers conducting their lessons, and testing the school's leading students. Although some made gestures towards testing the work of each student, these were shown to be hopelessly inadequate. One HMI, Frederick Cook, testified that it took four hours to inspect a school of 150 boys "very completely." Subsequent incredulous questioning of Cook revealed that what he *really* meant was an assessment of their "staple work" – hearing each student read aloud, seeing their writing, and testing their arithmetic – and that took an hour and a half. The remaining two and a half hours were spent looking at the condition of the school and administrative paperwork. Cook's figure was later derided as wildly unrealistic; it worked out to thirty seconds per student. Another HMI, Frederick Watkins, testified that in a school of 150 it took him at least four hours simply to assess all of the students. But even this figure meant that an HMI had less than two minutes to spend on each student, and *The Educational Times* wondered "whether an examination conducted at this railway pace can be worth much" at all.[70] Lingen was more forthright when asked if an inspector could assess whether *each individual* child could read, write, and "cipher" (perform basic arithmetic): "certainly not," he tartly responded.[71]

The Newcastle Report thus concluded that inspections were "of schools rather than of scholars, of the first class more than any other classes."[72] Lowe's insightful summary of the problem merits a longer quotation.

> An examination of classes is a very different thing from an examination of individuals ... Another objection to the reports of the Inspectors as a test of the actual efficiency of a school is their use of abstract phrases in describing the efficiency of the school. It is like the error in Platonic philosophy; they deal with the abstract and not with the concrete. They give a general notion of the schools, but they treat the school as something distinct from the scholars. They have examined a few children, and make a report, and, doubtless, a very true report, as to the quality of the education given to them. Their reports are full of such phrases as "the average proficiency of the children," but they speak of no particular child. They speak of the "general efficiency," the "general impression on the whole," "the general review," &c. They deal in impalpable essences, such as the moral atmosphere," the "tone," the "mental condition," not of the children, but as an abstract idea, of the school.[73]

The commission argued that only a quarter of the students in inspected schools could be counted as "successfully educated."[74] The implication was that the other three quarters were *unsuccessfully* educated. An article in *The Times*, possibly written by Lowe himself, claimed that "the great bulk" of graduating national school students knew little about "the plain subjects" of "reading, writing, and arithmetic" because these subjects were "undervalued and neglected."[75]

Some Newcastle commissioners were sceptical that the Education Department was drowning in paperwork, but the report did recommend that the department be relieved "of some of this mass of minutiae" in order to focus on larger "principles."[76] Nonetheless, it was a leap from concluding that school inspections and group assessments were insufficient to deciding that a school's funding should be tied to its students' individual examination performances – as the Newcastle Commission ultimately recommended. One historian has suggested that a majority of the commission members, led by James Fitzjames Stephen, were in favour of payment by results well before the end of the commission.[77] Regardless of why the Newcastle Commission favoured tying school funding to exam results, the DSA exam system of payment on results already seemed to show that such models worked. Lowe admired it, later describing the DSA system as "one of the most successful things that ever came under my notice" as Vice President.[78] On 9 May 1861 Lowe asked Cole to create a pass exam system in reading, writing, and arithmetic to which funding could be tied. By the end of that month Cole met with Donnelly and Kay-Shuttleworth to discuss the system. Kay-Shuttleworth was strongly critical and made numerous objections. Cole responded in a memo noting that DSA experience had worked well since 1857 in art teaching, then later in science, and Lowe was satisfied.[79] He gave £10 to Cole and Donnelly to conduct an experiment at a national school in Brompton, near DSA headquarters.[80]

In the spring of 1862 the Revised Code was formally introduced in the service of what we now call "legibility."[81] It was justified as a simple way to ascertain whether individual students at national schools learned basic knowledge and skills, as well as to make the education budget accountable to Parliament by making it "more simple, more general, and more effective." Cole's hand was shown in the Code's preference for "'piece' rather than 'day' work." Donnelly's principle of "elasticity" had moved to transform not only science education but nationally funded primary education as well.[82] Because the department estimated that the cost of schooling for each child was thirty shillings a year, the government would give no more than fifteen shillings per child; the rest of the money was supposed to come from tuition fees or local subscriptions, a belief which was never challenged.

Although the amounts changed over the years, the proportions remained about the same. The government paid for about half of the cost of a student's schooling. One-third of this amount came from the student's attendance, and two-thirds by examination passes. Of this latter proportion, 33 per cent was forfeited if a student failed to "satisfy the inspector" that he or she read to an assigned standard; 33 per cent if that student railed to write to that standard; 33 per cent if that student failed to perform arithmetic to it.[83] And the Revised Code did fulfil Gladstone's charge to Lowe to reduce the Education Department's budget: from £800,000 in 1861 it declined to £600,000 in 1865, even as new schools were added.[84]

The Revised Code set out four groups based primarily on standards of achievement and grouped more loosely by age. To pass reading, Standard I (ages 3–7) required the successful reading of a "narrative in monosyllables"; Standard II (ages 7–9) a short paragraph from an elementary reading book; Standard III (ages 9–11) a short paragraph from a more advanced reading book; and Standard IV (ages 11 "and upwards") a short ordinary paragraph in a newspaper. On passing each standard, the child was supposed to move up to the next one. What resulted was a severe restriction on what an HMI was free to do at each inspection, as procedures were also laid down about how he was to conduct each exam: whether slates and pencils or pen and ink were to be used, how his marking was to be done, and what kind of paperwork was to be submitted. More specific guidance about what constituted a pass was set out: a pass could be given if reading was not good but at least intelligible; if dictation was legible and most common words were spelled correctly; and if the student's arithmetical method was correct, getting at least one sum right.[85] The idiosyncratic standards of each HMI were being forced into more uniformity.

The best-known criticisms of the Revised Code are Matthew Arnold's. He complained that it turned HMIs into "registering clerks, with a mass of minute details to tabulate ... to ascertain the precise state of each individual scholar's reading, writing, and arithmetic ... It is as if the generals of an army ... were to have their duties limited to inspecting the men's cartouch-boxes [*sic*]."[86] Although Arnold's now-famous views on education were not always respected by his contemporaries,[87] his recognition of the new emphasis on tabulation was correct. In 1868 another critic noted that the Revised Code seems to have been drawn up by people "more accustomed to collecting and tabulating statistical data than educating"; it was "admirably adapted to ascertaining the exact amount of instruction imparted, and then tabulating this."[88]

Mechanical objectivity had arrived to schooling. The Revised Code's new standards – and the examinations sorting people against them – were informed by the desire to glean statistical information or abstract representations from what many saw as a system so complex that it evaded accountability. Arnold was above all distressed about the loss of the discretion to which individual HMIs had become accustomed, but if statistics are to be meaningful and useful, then the conduct of its collectors must also be standardized. National schools were being refashioned into places in which it was possible to collect useful statistics.

Ironically, Lowe later resigned his vice presidency over the charge that he was changing the reports of HMIs after they had been submitted – with the MP raising the charge being William Forster, Matthew Arnold's brother-in-law.[89] Yet the general principle of "payment by results" and its underlying principle of "elasticity" was not so easily dislodged. Later reformers tweaked the system in several iterations; new subjects like science and singing were included. Payment by results informed schooling policy until 1895.

We have seen how payment by results was reviled by later educationists as a grim and wasteful time in the history of schooling. Lowe would later say that the whole point of specifying *what* had to be known in each age group was not to limit what was taught but to make the learning examinable; in order to responsibly allocate public money, the conditions had to be explicit.[90] But he, and others like Cole and Donnelly, had not thought about the likely responses by teachers: told that *only* those topics were to be rewarded if demonstrated successfully, teachers would tend to focus *only* on those topics, teaching to those upcoming exams and ignoring everything else. Meanwhile HMIs, already under great time pressures, would tend to *only* assess reading, writing, and arithmetic and ignore other impressions that would not affect the grant. Such Procrusteanism had been foretold,[91] and these prophesies were borne out. Before 1862 some national schools had taught basic science in addition to the three R's, but upon passage of the Revised Code, this stopped. Schools also began to "prune" their student bodies, expelling, or preventing from enrolling, any student thought unlikely to pass the required exams.[92] In Chapter 7 we will go into more detail on attempts to game the system.

Teachers, students, and examinees, faced with increasingly rigid bureaucratic categories, sought to fit into those categories to their best advantage. In this, they showed agency and rationality: when faced with clear and explicit targets, focus on only such targets at the expense of all other subjects. Since the architects of the policy of payment by results were informed by deductive models of political economy, it is surprising they did not anticipate that the people affected by such policies would also follow these laws of political economy themselves.

3 An Epistemology of the Mundane: Dissecting One Examination

Whatever may be said about catechetical method in teaching, it had a demonstrable efficiency in the "assessment of performance," to use the modern jargon, so long as the questioner or examiner was satisfied by rote answers.

– Frank Foden, *The Examiner: James Booth and the Origin of Common Examinations*, 12

One May day in 1873, the notoriously busy T.H. Huxley would have gazed upon a stack of 6,834 Department of Science and Art examinations in animal physiology, knowing he had but three weeks to have them all marked and the scores tabulated and published. While he did not have to mark them all himself – he had at least six assistant examiners to help him – supervising this process must have occupied a considerable, and unpleasant, part of his year. Such examining activities tend to go unmentioned in histories about Huxley, probably because they are seen as boring in comparison to, say, the impromptu 1860 "debate" between Huxley and Samuel Wilberforce about human ancestry. Yet Huxley's examining activities were more historically significant than such evanescent discussions, because his exams were part of an emerging infrastructure through which large numbers of people could be accredited as knowledgeable in a topic like physiology. Such certification was formally recognized by the blue-green diploma of the DSA, but underlying this was a much larger and mundane system by which a credible expert attested that the holder of the diploma knew correct facts and definitions in a given scientific, technological, or mathematical subject. To better understand this process, and to grasp abstract questions about what "common" scientific knowledge is, we study the mundane infrastructure of a single examination from inception to completion – Huxley's May 1873 DSA examination in Subject XIV, animal physiology.

Preparing the DSA exam

T.H. Huxley had been an examiner with the DSA from the inception of its exams. He was in charge of two subjects: zoology, the exam for which was usually written by no more than a hundred candidates, and animal physiology, which by the early 1870s was written by at least 3,000 candidates. By 1873 a DSA examination consisted of three subexaminations, called "papers," in ascending levels of difficulty: elementary, advanced, and honours. The papers' requirements tended to be cumulative – questions from elementary papers might turn up on advanced or honours papers, and knowledge of facts asked on elementary papers was usually necessary to pass higher ones.[1] Once a candidate passed one paper, that candidate was certificated to receive payments for any student of theirs who subsequently passed that same paper. This point bears repetition: simply passing an exam in a topic gave that person the right to teach others that topic for the same exam, and to earn money for others' exam successes.

Candidates usually ranged from age eleven and up – some were aged forty-five – and girls and women had been writing DSA exams since 1861. It had been in 1872 that Margaret McConnish had placed highly enough on Huxley's animal physiology exam to win a scholarship to his South Kensington teaching laboratory.[2]

DSA exams were always written in May. Hence by the previous September Huxley, with the other DSA examiners, had written a syllabus for each exam, specifying the topics needed to be known for a pass; these syllabi were then typeset and distributed, along with various other rules and forms, as part of the DSA *Directory* mailed out to DSA certificate holders in October. By releasing a syllabus well before an exam, the administrators set a public standard that candidates needed to know and prepare for. This practice was also used by the Oxford Locals and Cambridge Locals. The University of London, military academies, and the Society of Arts went further and published past exam questions, which were then reprinted in popular textbooks such as Adolphe Ganot's *Physics*.[3]

For the animal physiology examination of 1873, Huxley's syllabus listed almost six pages of topics. Things necessary to know for the elementary paper took up two pages: some basic inorganic chemistry, anatomical names and facts, and the workings of specific physiological systems such as the nervous system. Topics necessary to know for the advanced paper added another two pages: understanding relationships between different physiological systems, or changes in those systems (e.g., blood coagulation; "unconscious cerebration"). The syllabus for the honours paper was deceptively simple: it might ask questions about any subject appearing in what Huxley called the "standard English works upon physiology": this included W.B. Carpenter's *Principles of Human Physiology* and his *Manual of Physiology*, and J. Marshall's *Outlines of Human and Comparative Physiology*; the suggested books had not changed since 1869.[4] Huxley was not above suggesting his own *Lessons of Elementary Physiology* (1868), published by MacMillan at four shillings and six pence. Since his textbook was highly regarded by teachers and exam coaches, such a tactic cannot have harmed its sales. In that same directory, we see Huxley's fellow DSA examiners also putting their own textbooks on the syllabi: in Tyndall's own acoustics, light and heat course, three of the eight recommended textbooks are his; in his magnetism and electricity course, one of two are his. Tyndall's friend Frankland, meanwhile, featured one Tyndall book in his own inorganic chemistry syllabus.[5] (In fairness, Tyndall's works were considered excellent.) Monetary considerations aside, putting one's own texts on syllabi is an excellent way to ensure that one's work actually gets read. It is probable that the most careful readers of Huxley's works were those studying for his zoology and animal physiology exams. Good coaches would not only tell their students to adopt this strategy, but emphasize topics that Huxley had historically asked about: his favourite questions. Such easy predictability was one reason why the DSA stopped recommending specific texts in 1876.[6]

Yet even ending such recommendations would not have been enough to make examinations less predictable and easier to game. Trustworthiness and accountability were usually the intentions of an institution when it employed highly visible and well-regarded scholars to draw up exams. Yet even advance knowledge of an examiner's identity could lead to coaches and candidates guessing about the nature of the questions. Huxley had to be predictable (as the aim was to induce candidates to study a certain list of topics), but not *too* predictable. Indeed, Huxley's identity as examiner was well known, and one teacher claimed that DSA exams displayed the most prominent "idiosyncrasies," or hobby horse topics, of any large external exam in the British Isles.[7]

Making things still more complex was the widespread knowledge in the coaching industry that some examiners simply set exam questions out of their own works. So one task for coaches and candidates was to read everything written by the examiners. For those taking Alexander Bain's paper in the moral sciences for the Indian Civil Service exam, "Magazine articles from his pen that bore even remotely on that subject were searched out and studied, and no means were left untried to adapt the candidate's reading to the known views of the examiner." Moreover, since Bain belonged to the "extreme sensational school" of psychology and was known for penalizing answers written from a different perspective (such as idealism),[8] savvy candidates knew the importance of adopting that same sensationalist perspective in any answer they wrote out.

A few months before the May exam date, Huxley sat down to write out the questions for the papers, assigning a numerical value to each question based on its importance and difficulty. One elementary question was:

> What is a gland? Mention any two different glands, and say where they are placed. (16 [marks])[9]

The advanced paper could only be taken by candidates who passed the elementary paper. In addition to requiring knowledge asked in elementary papers, advanced questions were wordier and required candidates to know additional specific physiological and anatomical terms.

> Describe the structure of the trachea, bronchi, and lungs. State why the lungs change their bulk in correspondence with the movements of the walls of the chest; give the volume of air inspired and expired at each respiratory movement; and compare the composition of the expired with that of the inspired air. (26 [marks])[10]

Again, only after passing the advanced paper could one write the honours paper. One honours question for the 1873 paper – the possible topics to study were simply listed as "any treated in the standard English works upon Physiology" – was, "Give an account of the development of the urinary and generative organs."[11] Such brevity is deceptive – much knowledge would have been needed to correctly answer this question, going beyond physiology into comparative anatomy and embryology.

The 1873 animal physiology exam had no practical questions.[12] It was argued that the theory papers were structured in a way that assessed if a candidate had done practical work, and inspector's visits to schools would show if practical work was being done.[13] Such a claim was a rationalization: there were administrative barriers to having candidates answer practical questions. Not only were practical exams more expensive to run; they would have been very difficult to standardize. Huxley's new teaching laboratories at South Kensington were intended to remedy the problem of learning science only from books by training new cadres of teachers at the bench.

A teacher stood to earn an average of thirteen shillings and seven pence for every student who wrote the DSA exams, about the same amount offered by the Revised Code for each student, but earning the DSA's money required substantially less teaching. As examiners, Huxley and his fellows were paid well by the DSA. For drawing up his questions, Huxley received ten guineas. But he also earned a hundred-guinea retainer, three guineas for each meeting, and £1 for every twenty papers marked. From available DSA rates I have calculated that Huxley earned about £160 for his work on the 1873 animal physiology exam.[14] By comparison, in 1867 the twenty-one examiners for the Cambridge Locals were paid only between £10 and £15 for several weeks' worth of marking, and in the late 1850s Tyndall refused to do any more examining for the Council of Military Education after they cut his fee in half.[15] Cole and Donnelly had reasoned that they needed the best possible people to organize the DSA exams, and wished to keep those people as examiners. Since examining was "of a very laborious and repulsive nature," it was important to pay well.[16] Such a position offered additional benefits to Huxley, such as a little more power to shape the future direction of the life sciences. He could direct some minor patronage of his own, nominating subexaminers who would be paid about £50 for their marking. Huxley's protégé Michael Foster was one subexaminer, and recommended some of his own Cambridge students for this work.[17]

The use of public money necessitated tight security. There are rumours of proofs leaking beforehand, possibly for medical exams, which created "a matter of considerable difficulty."[18] To prevent such leaks the questions and proofs travelled between Huxley and the printers in special envelopes. There was a list of specific firms to be used by the DSA for the printing of the exams, with one condition being their having a single locked room in which all composition, printing, and proofing was conducted. Eventually the DSA would create its own printing services at South Kensington, run by Royal Engineers. Finally, the proofs were watermarked so that a leaked proof could be traced back to a specific printer; only as many copies of an exam as were required were printed. By 1873, while there had been some minor cases of cheating here and there, there had been no major and systematic cases of examination fraud. Cole had attested to this before a parliamentary commission the previous year.[19]

Upon being printed off, the exams were placed in batches, in exactly the same number of papers as requested by each local exam centre, then placed into calico-lined security envelopes specially designed by Post Office experts. An amused witness described them as resembling Rob Roy's purse ("the simplicity of the contrivance to secure a furred pouch, which could have been ripped open without any attempt on the spring, &c").[20] Each envelope was glued shut, the flap sealed with melted candle wax and embossed with the official DSA seal, and popped in the mail. The package was supposed to be mailed directly to the residence of one of the neighbourhood committee members who administered its exam – preferably the committee's chair.[21]

Entering the House of Catechisms

One of the most stimulating parts of Andrew Warwick's *Masters of Theory* is his reconstruction of John Henry Poynting's answer to one of Baron Rayleigh's Mathematical Tripos questions in 1876. In so doing Warwick demonstrates how the examination paper acts as the closest thing to "real-time records of mathematical physics in the making."[22] This section reconstructs a far more mundane affair: how one candidate studied, memorized, and repeated definitions on a physiology examination. Catechisms – a series of questions and answers – were the primary form of learning and studying these definitions. By analysing this process, we may learn more about how common scientific definitions were memorized and used in an orthodox way. Both the exam and the candidate were more obscure than the Tripos and Rayleigh: the candidate for this May 1873 advanced paper of the animal physiology exam was Charles Ledger of St John's School, Woking.[23] Ledger had begun at St John's School as a pupil-teacher in May 1869, and he began taking DSA exams in the following year. By 1873 he had written papers in geology, acoustics, light and heat, magnetism and electricity, and inorganic chemistry.[24] Anyone was eligible to write a DSA exam, but only students of certificated teachers could earn money for their teacher; they also had to take a set number of lessons (in this case, twenty-four) in that subject, verified by a signed register open to inspection by visiting DSA officers. Ledger's teacher at St John's, Robert Goffin, was so certificated, and there was a signed register of Ledger taking the twenty-four classes, making Goffin eligible to earn money if Ledger passed.[25]

Ledger had to study for the exam. For simplicity let us focus on one topic that the syllabus warned might be on the exam: the trachea. Ledger's teacher, Goffin, had told him to focus on the trachea and to look specifically at Huxley's highly popular textbook for its definition, in Lesson IV, "Respiration": "Leading from the larynx downwards along the front part of the throat, where it may be very readily felt, is the *trachea,* or windpipe ... its walls are, in fact, strengthened by a series of cartilaginous hoops, whose hoops are incomplete behind, their ends being united only by muscle and membrane, where the trachea comes

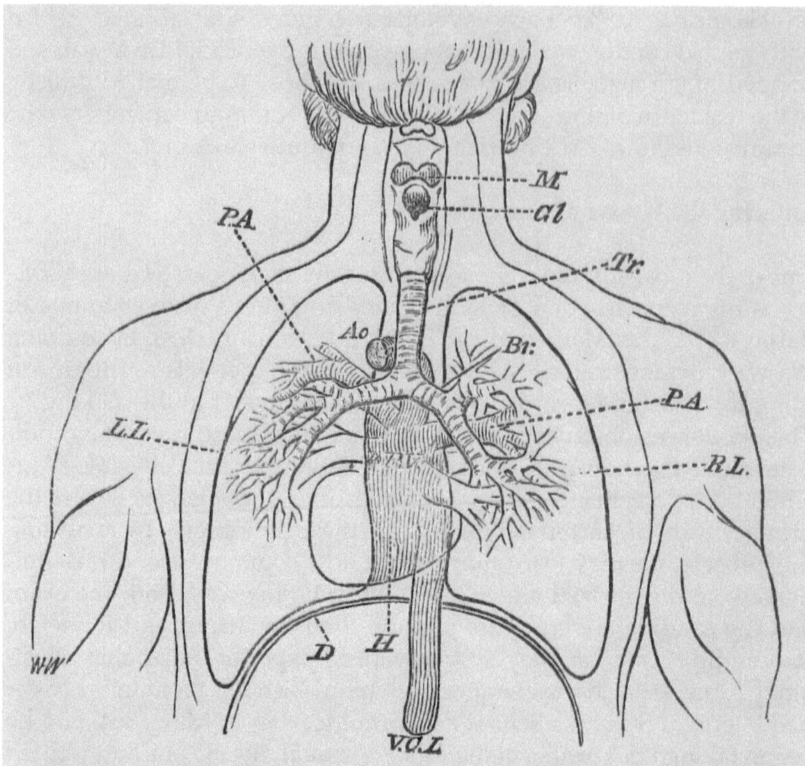

FIG. 19.—BACK VIEW OF THE NECK AND THORAX OF A HUMAN SUBJECT FROM WHICH THE VERTEBRAL COLUMN AND WHOLE POSTERIOR WALL OF THE CHEST ARE SUPPOSED TO BE REMOVED.

M. mouth; *Gl.* glottis; *Tr.* trachea; *L.L.* left lung; *R.L.* right lung; *Br.* bronchus; *P A.* pulmonary artery; *P.V.* pulmonary veins; *Ao.* aorta; *D.* diaphragm; *H.* heart; *V.C.I.* Vena cava inferior.

Figure 2. Cutaway diagram, human upper respiratory and digestive tracts, from T.H. Huxley, *Lessons in Elementary Physiology*, 2nd ed. (London: Macmillan, 1868), 90.

into contact with the gullet, or *oesophagus*."[26] The illustration from the textbook is shown in Figure 2.

More about the trachea also appeared in "cram-books," that is, books marketed as companion works to textbooks. One such work was Thomas Alcock's *Questions on Huxley's Lessons in Elementary Physiology for the Use of Schools*, published also by MacMillan in 1868. It was organized catechistically, and Ledger clearly memorized in this way.

Alcock's questions	Ledger's study notes
16. Describe the cartilaginous hoops of the trachea and of the bronchi, and state their use. 19. What is the size and character of the air-cells of the lung?	The trachea is a cartilaginous pipe through which the air passes into and out the lungs; at its upper extremity is the larynx; the lower extremity of the trachea passes into the thorax, and divides into two branches; these are called the bronchi. The lungs are composed of loose spongy texture, enclosing minute air cells surrounded by capillaries, and holding together the bronchi, which finally terminates in the trachea; there are three lobes to the right lung, and two to the left.
52. What are the ordinary twenty or thirty cubic inches of air taken in at each ordinary inspiration called?[27]	In ordinary breathing, 20 to 30 c. in., of what is conveniently called tidal air, pass in and out; it follows, after an ordinary inspiration, 100 + 100 + 30 = 230 c. in. may be contained in the lungs.[28]

One evening in May 1873 – unfortunately it has not been possible to ascertain the specific date – Ledger showed up at the neighbourhood DSA exam centre, in St John's School. Those exams were administered by a science committee headed by Theodore Wilks, vicar of Woking.[29] Ledger was likely there well in advance, as another of Goffin's notes to Ledger told him to "come early."[30] Upon arrival Ledger would have registered with one of the attending members of the organizing committee by giving his unique candidate number (used to identify candidates over multiple years of writing DSA exams), and by 6:50 would have been sitting quietly in his seat, positioned a minimum of five feet from the next candidate. Ledger would have brought in his own pens and pencils, but given standardized blank test booklets in which to write. At 6:55 the sealed exam package was opened by one of the committee members (for further trustworthiness they were requested to open it in the presence of two other members) and the printed questions were given to the candidates; for convenience all three papers' questions were in a single printed set, with the candidates then choosing the paper whose questions they would answer. There was a possibility that one of the specially appointed Royal Engineers might show up at that centre to ensure "uniformity of action" and proper implementation of the DSA's rules; it is unclear if a Royal Engineer visited that evening, but we do know that those assigned to the Surrey district that May were Major E.H. Courtney and Captain Warren Wynne (who together visited twenty-three sites in Surrey and Berkshire that month).[31]

At seven o'clock, Ledger and 6,833 other candidates across the British Isles began writing their Subject XIV papers; they had three hours in which to complete them. Ledger was one of 1,491 candidates to take the advanced paper.[32] Lacking a record of what went through Ledger's head as he wrote the answers, we might use Trollope's wry description of a difficult examination in its place:

> A man has before him a string of questions, and he looks painfully down at them, from question to question, searching for some allusion to that special knowledge which he has within him. He too often finds that no such allusion is made. It appears that the [examiner] of the occasion has exactly known the blank spots of his mind and fitted them all.

The anxiety of the candidate is heightened upon seeing one's neighbour scribbling away quickly, "as though reams of folio could hardly contain all the knowledge which he is able to pour forth."[33] Conversely, for a more successful experience we can relate an account from a candidate for the 1867 Maths Tripos.

> It was with a curious sensation that I opened the paper the examiner gave me and read the first question ... I felt that I knew the requisite figure, but except a kind of instinct that if I only began it must come out, I was conscious that I had not the remotest idea what the steps of the proof were. This feeling soon wore off, my mind cleared itself, and I was not conscious of anything around me but a general confused sensation of scratching pens ... Then to work again until the first stroke of twelve, when an examiner close by startles me by calling out sharply "No more writing.[34]

Ledger's experience was probably closer to the second, more successful account. For Huxley did indeed ask a question about the trachea.

Huxley's question	Ledger's answer
[Question 21]. Describe the structure of the trachea, bronchi, and lungs. State why the lungs change their bulk in correspondence with the movements of the walls of the chest; give the volume of air inspired and expired at each respiratory movement; and compare the composition of the expired with that of the inspired air. (26 [marks])	The walls of the trachea are strengthened by a series of cartilaginous hoops, which hoops are incomplete behind, their ends being united only by muscle and membrane, where the trachea comes into contact with the oesophagus.

Huxley's question	Ledger's answer
	The trachea passes into the thorax and divides into two branches; these are called the bronchi.
	The lungs are composed of loose spongy texture, enclosing minute air cells surrounded by capillaries, and holding together the bronchi, which finally terminated in the trachea, three lobes in the right and two in the left.
	The trachea is a cartilaginous and membranous pipe through which the air passes into and out the lungs; at its upper extremity is the larynx....
	In ordinary breathing, 20 to 30 C. in., of what is conveniently called tidal air, pass in and out; it follows, after an ordinary inspiration, 100 + 100 + 30 = 230 C. in. may be contained in the lungs.
	The expired air contains (CO2) carbonic acid, and the inspired air contains oxygen, they therefore exchange the blood, taking up the oxygen and giving up CO2.[35]

By ten o'clock the exam ended. All booklets were collected and placed in a security envelope along with another form from the organizing committee attesting that all DSA exam rules had been followed; the envelope was then mailed back to South Kensington. It has not been possible to further reconstruct Ledger's other answers, although we do know he failed this particular exam, likely because his teacher was suspected of cheating. Was Ledger's answer a little too good?[36] We will return to him in Chapter 8.

Securing Uniformity

Given the speed of the Victorian post office, the completed exams likely showed up from the 832 different exam centres across the British Isles the very next day.[37] Now Huxley had three weeks to ensure all 6,834 exams were marked. He would have had six assistant examiners to help him: one for every 1,000 papers. Donnelly had arrived at this figure (and the three-week turnaround time) by calculating the rate of marking to its tolerable limit – seventy papers marked in a six-hour day. For assistants Huxley had hired men he deemed "able and distinguished" in physiology, since marking required quick yet accurate judgments. Each assistant earned about £50, a figure set using Cole's preferred remunerative principle of piecework: £1 for every twenty elementary papers marked, and £1 and ten shillings for twenty advanced papers. Huxley marked all honours papers himself.[38]

Before anyone marked a single paper, Huxley convened a meeting with all of the assistant examiners. They established what constituted first- and second-class answers to each question. As we lack specific information for each answer, Donnelly's suggestions in his memos will suffice: first-class papers should "be within the reach of a well-taught, clever, first year student, or an average second year student," perhaps a score above 69 per cent in the elementary paper and above 74 per cent in the advanced paper. This meant about one-third of the elementary papers and one-quarter of the advanced papers should receive a first-class mark. A second-class paper should be open to "a fair, even a large, proportion of moderately stupid students" taught well for thirty to forty lessons: say, over a 29 per cent on the elementary paper and 39 per cent on the advanced paper. The answers on the advanced paper could be answered by the average medical student. The honours exam Donnelly tersely suggested ought to be "very stiff" – Huxley could mark it as he wished.[39]

At the end of this first meeting each assistant carried away his batch of 1,000 papers. Huxley would mark any remaining papers, in addition to the honours ones. To ensure further standardization, Huxley also had to look over at least 20 per cent of his assistants' papers too, which in 1873 would have worked out to 1,366. A week later, there was a second meeting to again standardize marking; assistant examiners would each bring with them fifty to sixty representative or anomalous papers; they compared them, asked Huxley questions, and revised their marks accordingly.[40]

After three weeks, the marked examinations were returned to South Kensington; figures were given to DSA clerks, who had a week to tabulate the results, compare them against the lists, and publish them. The chief examiners might hold a general conference at which they swapped suggestions and tips.[41] Upon viewing the results, a certificated teacher would send in a form to request payment. If Ledger received a first-class result on his advanced paper, Goffin would earn £2; he would earn £1 if Ledger got a second class. (The same rates applied for elementary papers, while a first class on an honours paper paid £4). Although students themselves earned no money, they could win medals, books, or instruments as well as scholarships to Huxley's South Kensington teaching laboratory. The May 1873 exams yielded Goffin £49 from St John's School candidates and another £51 from his teaching of candidates at another school in nearby Guildford. He also taught the gold [Queen's] medallist in geology that year as well.[42] Reverend Wilks's testimonial for Goffin noted that "the remarkable success of [Goffin's] pupils shows that he has the capability of efficiently teaching what he knows."[43]

One pitying observer remarked that for those marking the DSA exams, "to look over a thousand sets of answers to the same paper by people you don't care about, is next door to penal servitude,"[44] and it's hard not to shudder in agreement. Yet it is also difficult not to be impressed by the work that went into organizing these exams. The system governing them was especially rigorous and highly formalized because of the amounts of money at stake: this induced the examiners not only to try to be impartial but also to trumpet this impartiality. The ultimate consequence of this work meant everyone was able to ignore the form of the exam and the railway tracks on which it ran, and focus only upon its content. Success meant invisibility.

Common Knowledge

DSA policy was to destroy all completed scripts after one year, so we must speculate how Ledger's answer about the trachea was marked by Huxley or one of his assistant examiners. This has been complicated by the fact that Ledger failed the larger exam, probably because his teacher was accused of secretly learning, then teaching, the exam questions in advance.[45] This is why Ledger's definition of the trachea is so close to Huxley's textbook definition. In other circumstances, however, one can see how catechetical knowledge tends to be rewarded on examinations. Such answers do tend to be easier to mark, because they show clearly whether someone knows or does not know some discrete item of knowledge.[46]

A catechetical format not only rewarded the correct answering of questions about definitions, facts, or principles; it could also reward answers that agreed with a particular *perspective* held by the examiner. Recall Bain's habit of rewarding anyone giving answers from his own sensationalist perspective and punishing anyone taking a different one. On Huxley's animal physiology exam, most of the questions required no particular philosophical perspective – the correct answers could be memorized from textbooks. Thus despite

Huxley's reputation for controversy, on his DSA physiology examinations there were few questions that could be answered from his favoured materialist perspective, such as life being based in protoplasm or consciousness being simply a matter of compounded reflex functions.

Nonetheless, there were some questions whose answers did depend upon one's philosophical premises. For instance, another question on Huxley's honours paper in 1873 was

> 41. Give an account of the structure, physiological importance, and development, of the cornea, iris, and crystalline lens.[47]

As with an answer about the trachea, any answer about the eye's anatomy, physiology, and development would probably be simply a matter for a candidate to memorize and write out. But if the candidate decided to go further and *explain* why the eye's anatomy took such a form, a choice presented itself. Had such an exam been set before 1859, no other explanation would have been possible except through the language of natural theology – that is, William Paley's explanation that God made the "crystalline" lens of the human eye thinner to better focus on light refracted through the medium of air. This was in contrast with the thicker lens of a fish eye, better suited for light's refraction through water. But since the question was asked fourteen years after the publication of Charles Darwin's *Origin of Species*, even the best-prepared candidate who used natural theology to answer the question would have been seen in a dimmer light by Huxley. A milder case, in 1870, saw honours candidates being asked to explain not only the anatomy and histology of the spinal cord but also its "functions so far as they have been determined by experiment." Any candidate who answered *without* discussing the reflex arc – which, after all, had been introduced as a purely mechanical function by Marshall Hall, controversially undermining notions of a unitary soul, single consciousness, and free will – would not get full marks.[48] If Huxley was anything like Bain, he would penalize such an answer. This candidate might do very well on other questions, but a poor result on a single question could hinder the candidate's chance for advancing to further study in science (for instance, not earning a scholarship to attend Huxley's South Kensington laboratory). In this quiet way, Huxley and other examiners could reinforce their preferred perspectives, such as a worldview that was more naturalistic.

This aspect of the exams, along with their efficiency, helps explain why in 1861 Huxley described the department's exams as among the most important measures ever taken to reduce what he called "Parsonic influence" in schools.[49] As previously mentioned, Huxley's most immediately influential writings were probably not his articles or books but his textbooks and syllabi. The latter reached a larger audience: teachers' livelihoods depended on knowing the anatomical and physiological facts Huxley's textbook listed and instructing students to frame their answers in a manner acceptable to an examiner like Huxley. In 1873, 6,834 people wrote the exams in animal physiology, but two to three times that number studied for the exam and did not take it.[50] Thus we can conservatively estimate that in 1873 at least 13,000 people closely followed and even memorized, as Ledger had, whatever Huxley wrote about animal physiology.

From this consideration of mundane repertoires and paperwork we can move toward more abstract concerns. One is how scientific knowledge comes to be held in common – how one can be confident that others know the same things as oneself. Safely driving in traffic depends on each driver not only knowing the convention that a red light means stop and green means go, but also being confident that all other drivers on the road know and follow that same convention.[51] When one uses the word "trachea," then, how can one be confident that *others* also understand it to be a series of cartilaginous hoops? While it is difficult to imagine someone coming up with an alternative definition for trachea, the question becomes more important for more complex definitions. How about "Darwinism"? For the 1909 celebration of the centenary of Darwin's birth, eminent biologists showed up in Cambridge and invoked at least three different versions of Darwinism. While all agreed that it meant descent with modification (phylogeny), they differed on precisely what caused phylogeny: mutation (Hugo de Vries); acquired characteristics, later called "Lamarckian" inheritance (Ernst Haeckel); or strict natural selectionism (Alfred Russel Wallace). All three claimed that were Darwin still alive he would have supported their view – despite two mechanisms (Lamarckism and strict selectionism) being utterly in opposition.[52] Although the DSA exams were no longer being written by 1909, how might a student of the time have defined "Darwinism" on a test – and received a passing grade for that answer?

Wallace, co-discoverer with Darwin of the mechanism of natural selection, famed naturalist and biogeographer, can help us think about this issue some more. For just as Huxley was administering his 1873 animal physiology examination papers, Wallace was himself marking his quota of 1,000 papers for physiography (physical geography), DSA Subject XXIII. He had been marking since 1870, ever since his friend Henry Walter Bates alerted him to this chance to make some extra money.[53] Despite his renown, Wallace was only an assistant examiner, simply marking rather than drawing up the questions; geographer David Thomas Ansted was the official examiner.

The physiography exam was usually the most popular of all the DSA exams, particularly favoured by female candidates. In 1873 there were 15,238 candidates, almost three times as many as those who took animal physiology.[54] As he marked, Wallace amused himself as so many markers do – recording incorrect and inadvertently humorous answers. This was to contribute a "little hilarity" in his three meetings with the other examiners. To the question "What is meant by the distribution of plants and animals in vertical and horizontal space, and what do you understand by representative forms?" he recorded such answers as

- Horizontal distribution is when they grow near the horizon; vertical distribution is when they grow in vertical space (such as wheat).
- Plants grow in gardens, animals live on the earth.
- Representative forms of animals and plants is, how they are represented in books.

As still occurs today, some candidates stumped by a question had tried to answer by recycling it, defining each of the important words as best they could. In his autobiography, Wallace complained that such replies showed "gross ignorance of the facts" of physical geography, and cited other humorous answers from the physiography exam over the years: The Atlantic Ocean was 90,000,000 miles deep. Animals going extinct since humans appeared on the earth included giants, Jonah's great fish, unicorns, and *Productus horridus*.[55]

In the case of unicorns, the depth of the Atlantic, or "horizontal distribution," it is hard to disagree with Wallace's description of those answers as ignorant. Yet since we cannot simply fall back as an explanation on what we now know to be correct,[56] we can try to understand other answers with more charity than Wallace did. When it comes to what "representative forms" means, the question seems to have been testing whether someone knew the definition specific to physical geography: Ansted's 1868 textbook defines "representative species" as "Species of plants or animals in distant countries having similar climates which perform the same or a corresponding part in nature. Thus the puma of South America represents the lion of the old world, and the llama the camel."[57] However, if one reads the answer that representative forms of organisms are those portrayed in books, it is possible to interpret the answer as correct, if obtuse (in that the descriptions or images of organisms in books literally *are* representations). With extreme charity, one can see certain answers Wallace deemed "ignorant" as being correct in a different context. With the answer that "geographical range" meant that organisms were "arranged according to their shape and size," the candidate possibly mistook "geography" for some kind of taxonomic arrangement. With another candidate's answer that "geographical range" meant a place in which no species lived except the one "which first originated there,"[58] a possible reading (again, being so charitable as to come close to Bertrand Russell's warning about being too open-minded) is that the candidate might have been thinking about some form of evolution.

Wallace presumably felt that someone competent in physical geography ought to know the accepted definitions and so marked the above answers as incorrect. With Ansted and other assistant examiners all bestowing or withholding marks, they reinforced, in miniscule ways, the acceptable and unacceptable uses of various words in physical geography. Such a marking process rewarded the correct knowledge of scientific facts and reinforced some conformity to definitions accepted by a community. The cumulative result of these tens of thousands of tiny rewards here, and rejections there, helped ensure the survival of accepted, orthodox definitions while halting the spread of heterodox ones.

Following the September 1872 syllabus asking students to know about plant and animal distribution, the relationship between horizontal and vertical distribution, and the meaning of representative species,[59] Ansted drew up the 1873 physiography exam to test candidates' knowledge of these necessary definitions. Candidates answered as best they could, and then people like Wallace marked their answers. Might we see this general process of examination – the creation of questions, the answering of those questions, the marking of those answers, the transmission of those results – as a routinized communication circuit? The idea comes out of arguments that histories of science ought to study knowledge "in transit" as it circulates between people – especially focusing on regular patterns of such knowledge circulation.[60] Histories of how readers interpret texts have shown that different people may interpret a book in extremely different ways than its author intended.[61] Perhaps, then, there is an analogy between an examiner as an "author" and an examination candidate as a "reader." In this light, candidates who claimed that "representative types" meant depictions of a species in a book, or that "geographical range" meant taxonomy or phylogeny, were coming up with different definitions that in another world might be logically possible.

Yet if the general process of examination is indeed a routinized form of communication, it is an unusual one, for few communications circuits end up cycling back to the original author of a text. That is, if we use the analogy of examiner as author and candidate as reader, it is rare for an author to check and see if their readers' various interpretations of the message are the same as the author's. Authors usually don't get the chance to declare whether a reader has gotten something right or wrong. But such "diverse" understandings of scientific, technological, and mathematical facts, definitions, and principles were made more uniform when examiners had the chance to proclaim some correct and others incorrect – or, to be more precise, a candidate demonstrating knowledge of facts, definitions, and principles that agreed with what was accepted in the larger scientific community could pass on to a subsequent stage; those who did not demonstrate such knowledge could not. This process led to a subtle reinforcement of scientific orthodoxy, or – to be less pejorative – *common knowledge,* and large-scale exams like the DSA's helped to spread such definitions widely.

PART TWO

Examiners

4 Daguerreotypes of the Mind: Paper, Partition, and Specialization

When the oral method is adopted, none but those personally present at the examination can have any accurate or valuable idea of the appearance of the school ... Not so, however, when the examination is by printed questions and written answers. A transcript, a sort of Daguerreotype likeness, as it were, of the state and condition of the pupils' minds, is taken and carried away, for general inspection.

– Horace Mann, "Boston Grammar and Writing Schools," 334

"Where does the trachea commence and end?" This question was asked, aloud, by one of the ten members of the Court of Examiners at a Royal College of Surgeons' examination in the early 1830s. The candidate's audience of examiners sat around a horseshoe-shaped table; it was probably Friday night between 6 p.m. and 1 a.m., and while the examination usually took only between fifteen and thirty minutes, the examiners had probably been sitting there for some time. This is because the candidate for membership would answer the question aloud in what was known as a viva voce – living voice – examination. We lack a record of his answer, but we know that the next question was, "What prevents a piece of food getting into the larynx?"

The questions seem to have been designed to elicit the candidate's knowledge in a guided sequence in order to determine if he knew specific anatomical relationships and accepted surgical procedures. Thus questions immediately preceding the ones about the trachea and larynx were, in sequence:

Have you ever dissected?
Suppose you take off the skin of the neck, what do you first see?
What nerves and veins accompany the carotid artery?

Describe the mode of tying it.
The course and branches of the *par vagum*?
How are the constrictors of the pharynx one with respect to the other?[1]

The question about dissection led to one about the neck and its underlying nerves and blood vessels, then to the carotid artery. This was followed by questions about what are now called the vagus nerves, which run through the neck and are in contact with the pharynx, which touches the windpipe, which brings us to the trachea.

Because this was an oral exam, the examiners had some discretion to linger on one area, particularly if the candidate seemed to lack knowledge in it. This flexibility is one obvious difference between an oral exam and a written one, although there are different kinds of oral examinations. One kind, the disputation, was a ritual debate in which a senior scholar would take a position (e.g., criticizing an author's stance) and the candidate took the opposite (defending that author's stance); both sides sought out fallacies or weaknesses in the other's argument. Sometimes this took hours. Another kind of examination was similar to the Royal College of Surgeons exam discussed above but had an audience, making it "public." An audience brought with it greater scrutiny of the candidate and examiners, but also the potential for drama: one or both sides might be embarrassed, or there could be inappropriate, insulting, or even politically dangerous remarks made, often in the form of puns or allusions.

Viva Voce Exams

Other scholars have already noted what characteristics were rewarded by oral examinations: wit, politeness, and memory. Andrew Warwick has pointed out that these exams favoured traits useful for a life spent in public: assertiveness, gentlemanly manners, the ability to think on one's feet and wittily respond to allusions. Even in what is now popularly viewed as a solitary pursuit – mathematics – such skills were valued in Enlightenment-era Cambridge, where disputations on propositions in Isaac Newton's *Principia Mathematica* were often conducted.[2] Warwick and Christopher Stray have shown that once written examinations came to dominate, other traits rose in importance, such as inner cogitation and even clear handwriting. Two characteristics of oral exams are of interest to us here: the great discretion possessed by an examiner asking questions orally, and the difficulty of seeing an oral exam as anything other than a whole performance.

Discretion is used here to mean the freedom to make decisions using one's judgment.[3] At viva voce exams, an examiner had the freedom to respond to specific circumstances: to skip a topic the candidate obviously knew, or conversely, to ask additional questions to probe a weak answer. Proponents of viva voce exams also claimed that an examiner was better able to distinguish between a candidate's ignorance or simple fear. Such on-the-spot flexibility, however, could easily turn into unfairness if the questions asked differed too much between individual candidates. This was a charge periodically made against the Royal College of Surgeons by authors in the reforming medical journal, *The Lancet*: sometimes a candidate received far easier questions than others in the same group, and the opposite also occurred. In one documented case, the candidate failed because he was asked outrageously difficult questions not given to others.[4]

Even when viva voce examiners did their utmost to ask the same questions of each candidate, the standards as to what constituted a satisfactory answer could still vary widely. Thus one guidebook for Oxford students warned of three distinct types of examiners they might encounter there. The preferred version was the kindly one seeking to both assess and instruct the candidate, but students often encountered either the jolly but indifferent examiner or, worst of all, the examiner who deliberately sought to frighten candidates.[5] If such diversity among examiners did not pose enough of a problem for uniformity and fairness, the mathematician Isaac Todhunter pointed out that a single examiner's *own standards* tended to fluctuate over the day as different candidates passed under his gaze.[6] This made assessment dependent upon moods, propinquity to meals (an ideal time for one's exam to be marked is immediately after lunch), or the performance of the previous candidate (it is difficult not to assess an answer relative to the performance of the previous candidate).

Such issues indicate a second characteristic of viva voce exams of relevance to us: the tendency for an oral exam to be seen as a single, whole performance. In other words, the evanescent medium of questions asked and answers given aloud made it harder for both examiners and candidates to distinguish between specific components of an exam. Oral examinations were difficult to break into sections. There is also the matter of the candidate's performance: it can be difficult to distinguish the substance of a candidate's responses from his or her posture and manner of speaking. In 1858 one London student magazine complained about well-prepared yet nervous medical students failing their viva voce exams, while "bold as brass" students bluffed their way through.[7]

Moreover, examiners might be overly influenced by a candidate's background. In June 1849, when the future chemist Henry Roscoe took London's Matriculation exam, while he translated Latin passages from the *Aeneid* on paper, the examiner moved around the room and asked additional viva voce questions of each candidate. Finally he came to Roscoe. After he made "rather a hash" of genders and tenses the examiner smiled grimly, and asked Roscoe if he was related to Henry Roscoe the historian. After affirming this, the examiner said, "that'll do, sir," and moved to ask questions of the next candidate. Roscoe passed first class.[8]

When compared with paper exams, then, viva voce exams were often seen as wanting. At Cambridge, on such examinations as the Mathematical Tripos, there was increasing pressure to make careful distinctions between candidates, particularly when social status and valuable prizes were at stake; as we shall see below, paper examinations made it easier to make distinctions that were seen as fair. By contrast, when visiting Oxford in 1845 the economist Richard Jones wrote to William Whewell about the unfairness of viva voce exams for ranking individual candidates, given how closely candidates might appear to one another and the wide variations between individual examiners.[9] As both Jones and Whewell were from Cambridge, one might expect some prejudice, but another observer of Oxford's viva voce process, seemingly one of its own graduates, noted "more disappointment caused by this branch than all others put together."[10] In 1839, a report to the University of London senate on medical education had gone further, arguing that it was impossible to rank students by order of merit using only a viva voce exam.[11] If a new epistemic ideal of the mid-century was objectivity – the attempt to separate as far as possible one's own judgment from what one was observing[12] – then viva voce exams did not fulfil this ideal.

Yet it is important to note what viva voce exams can do that written ones cannot. Oral questioning, done sensitively, can catch bodily details – a slight pause, a facial expression that merits more probing, other glimpses of a candidate's *habitus*. Oral exams were used especially for languages, but also for catching misconduct – in all the exam systems I have studied, examiners reserved the right to oral questioning if necessary. Yet this right seems to have been exercised rarely.

Partition

This chapter's epigraph begins with the American educational reformer Horace Mann's claim that paper examinations could form a "daguerreotype likeness" of pupils' minds, providing a "permanent record."[13] Mann's use of paper examinations to test secondary students was revolutionary for the time.[14] The contrast between oral and written communications is so taken for granted that it merits further exploration.

The anthropologist Jack Goody points out that setting a communication in writing made dynamic utterances more permanent, which facilitated later scrutiny; one could not only store a written communication but also compare it with other, similar records, engaging in what Goody calls "critical scrutiny." In his meditations on what he calls "inscriptions," Bruno Latour makes a similar point.[15] We possess the Royal College of Surgeons' exam questions that began this chapter because someone remembered them, wrote them down, and sent them to the 1834 Parliamentary Select Committee on Medical Education. Such new scrutiny placed the then-President of the college on the defensive when he testified before the committee as a witness. After complimenting the memory of whoever had forwarded the questions, George Guthrie continued, slightly menacingly: "If the Committee will be pleased to call up the gentleman who furnished the paper, and to allow me to examine him before the Committee, I would convince them, out of that paper, that he had got enough."[16] Guthrie sought to return the questions to the familiar oral context in which they had been asked – and from which they had escaped. Recall, meanwhile, Cole's near-contemporaneous push for written records that actually matched the testimony of a witness. People who appreciated the flexibility of the spoken word must have felt the new tyranny of the written word: more publicity and openness, to be sure, but also less discretion and less trust – the trust that a gentleman thought he had the right to expect.

Paper examinations facilitate the separation of an answer from the personal and physical context in which it is given: from the candidate, from the setting, from the time of answering. We have already seen Mann's, Goody's, and Latour's observations on how paper can freeze actions and separate an answer from its immediate context. This separation is called *partition*. Sociologist Elihu Gerson notes how partition breaks down a complicated process into multiple simpler tasks. Each task is seen as independent of the other, becoming a member of a set rather than part of a larger whole.[17] Partition seems to be the same process that makes possible what the philosopher Yaron Ezrahi defines as accountability. Tasks can only be assessed against general standards by being seen not only as specific acts, but also as performances detached from a person and his or her "inaccessible subjective dimensions." Such detachment facilitates the exposure of each task or performance to external scrutiny and comparison against neutral standards.[18] In short, an act is easier to judge when partitioned from a person and from other acts, and a paper record makes this partitioning easier.

Compared with viva voce examinations, paper examinations facilitate partition in three ways: by space, by time, and by organization. *Spatial* partition is the most obvious way in which exam answers can be separated from the place in which they are given. Mann's description of written exams as providing a "daguerreotype likeness" of pupils' minds meant that their answers could be taken away from the schoolroom, or wherever the answers were given, and looked at in a different setting. Spatial partition occurred when Guthrie's questions were transported and revisited by a later parliamentary committee; in 1846 when teenage boys in Sydney took the London Matriculation exam; or in 1857 when other teenage boys took the Oxford Local exam. In the latter instances, candidates answered questions that had been created from afar; their completed scripts were placed in packages and returned to the faraway examiners for marking. Indeed, written exams made possible even finer forms of partition within a single exam. The grammar school teacher R.H. Quick had students answer each question on its own sheet of paper; when the exam was finished, they signed their names at the top and placed each answer-sheet in its own pile with others' answers to that same question. In so doing, Quick sought to solve the problem raised by Todhunter – that one's own standards might fluctuate during the day. By marking all the answers to a single question against a "fixed standard" in his mind, Quick had created a simple form of monitoring his actions – a humble *aide-memoire* to better ensure fair marking.[19] In modern parlance, we might even say that Quick had externalized some of his cognition.[20]

Temporal partition is made possible by spatial partition. It means that a question or answer can be revisited later, be it a candidate deciding to leave a difficult question until the end of an exam, or an examiner deferring judgment for a moment. For instance, it was only because the questions for a University of London 1844 physiology exam were written down that other examiners could, after the exam had been conducted, learn that the exam actually had no physiology questions.[21] The change from oral to written questions was marked by a greater ability to distinguish between the different components of an exam. When in 1709 Richard Bentley, as Master of Trinity College, Cambridge, introduced a written essay to allow more time to assess fellowship candidates, putting the answers on paper enabled both spatial and temporal partition: they could be taken away and read at the examiners' convenience, where before their judgments had to be made closer to the examination event itself.

The extraordinary demands of the Mathematical Tripos were facilitated by spatial and temporal partition. By 1827 the Tripos questions were no longer read aloud but printed in advance for "stability." Spatial and temporal partition made it possible in 1836 for an examiner to mark each question individually rather than relying on his general "impression" of the candidate's answers, as had been the case previously.[22] When a question for the Tripos was drawn up by one examiner, the other examiners carefully studied it and sought with the "utmost severity" to find fault with it;[23] this was spatial and temporal partition put to work in an adversarial context.

Organizational partition of exams is made possible by spatial and temporal partition. For Gerson this can then be combined with standardization, a double move taken up at length in the next chapter. A group of questions or a group of answers can be broken into individual questions, then placed into new groups deemed to better reflect some underlying common property. The Royal College of Surgeons' questions at the beginning of the chapter seem to have been organized according to body part proximity. It is as though the candidate's gaze is led from part to neighbouring part. The order of questions moves from the skin of the neck to the carotid artery, and then to the vagus nerve which sits next to that carotid artery. But when one breaks up the questions, seeing each as a separate entity, different organizing principles emerge. For instance one might group the questions by anatomy ("Where does the trachea commence and end?" "[What are] the course and branches of the *par vagum*?"). Or one might group them by familiarity with certain techniques ("Have you ever dissected?" "Describe the mode of tying [the carotid artery]."). Such regrouping facilitates more discussion of discrete topics, such as anatomy or surgical technique – one set of questions on anatomy, and one set on the practice of surgery and dissection.

Organizational partition also makes greater specialization possible. Indeed, this was what the mathematician Charles Babbage pointed out in his 1832 *Economy of Machinery and Manufactures*. To illustrate the "division of labour," Babbage cited Adam Smith's famous case of the pin manufacturer. A single pin could be made by a single labourer who performed each of the required steps: drawing a wire, straightening it, cutting that wire, sharpening its point, and so on. What resulted was a small number of pins, largely because the worker had to switch between multiple tasks. But when one partitioned the making of a pin into its constituent tasks, one worker was responsible for one specialized task: one to only straighten the wire, one to only sharpen the point, and so on. For Babbage, in addition to no longer having to waste time switching between tasks, each worker could gain proficiency at this single job and become more efficient. Machines, too, could be built that would perform these single repetitive tasks far more quickly than a human worker could.[24]

It is these forms of partition that separate performances from persons. The viva voce examiner found it difficult to distinguish between Henry Roscoe, the examinee making mistakes in Latin declension, and Henry Roscoe, son of the historian. On orally examining medical students, it could be difficult to distinguish between the confidence of the candidates and the often-hollow substance of the answers. When an examiner was faced with two very evenly matched candidates at Oxford, it was difficult to make careful distinctions between them on the basis of their oral answers alone. To be sure, it was still possible to continue with viva voce questions and make careful distinctions: the Royal College of Surgeons did not adopt paper examinations until 1860. Yet in reaction to various complaints about its examination procedures, by the 1840s it had divided the Court of Examiners into four groups. Each group of examiners (two or three sitting at a table) focused on a specific subject, each rendering their own judgment on whether the candidate was knowledgeable about that subject. And it made greater efforts to better distinguish candidates' knowledge: "good," "moderate," and "bad."[25] Even though rudimentary, this was partition at work, showing that paper facilitates, though is not strictly necessary, for partition to occur.

Mechanization

If we return to specialization, Babbage also pointed out that the division of labour could be applied to complicated projects requiring great thought. A "division of mental labour" was possible if one partitioned complex projects into sets of simpler tasks. Babbage gave the example of a French savant who, during the Revolution, was charged with the creation of seventeen volumes of mathematical tables. The savant decided that the most efficient way to proceed was with three groups of workers: sixty to eighty of them needed to know nothing but addition and subtraction; seven or eight had to be well acquainted with mathematics; and five or six of them had to be the best mathematicians in France. Despite requiring the least knowledge, the first group still performed the most arithmetical operations; the second group checked the work of this first group; and the third and most skilled group oversaw the operations of the first two and directed the project as a whole.[26] This first unskilled group Babbage called "mechanical": their tasks were simple and repetitive, requiring little mathematical knowledge or the need to exercise judgment.[27] For Babbage, such basic work could be performed "mechanically" whether by an actual machine or by a less-skilled person; this is why Babbage wished to mechanize routine arithmetical calculations with his Difference Engine. The first "computers," after all, were not machines but people.[28]

Jon Agar's *The Government Machine* takes up the metaphor of "mechanical" and "machine" and shows how this image went on to shape the civil service. Its reformers, like Trevelyan, created a division between the "intellectual" and the "mechanical" workers. While it is unclear if he was directly inspired by Babbage's notions, Trevelyan's organizational scheme certainly resembles them. Hence the intellectual role was to be filled by university-educated civil servants who deployed gentlemanly discretion; the mechanical role was to be filled by clerks who performed such repetitive tasks as copying and thus had no need to exercise much discretion, if any. Trevelyan thought this reform would greatly increase the efficiency of the civil service. A collection of mechanical clerks would act like parts of a machine – in that they would be interchangeable. Ideally, one could swap one clerk in for another, and there would be little difference in the work accomplished.[29] Yet as with the lowest group of workers on the French mathematical tables, it was this "mechanical group" that performed most of the work necessary for the smooth running of government.

To work "mechanically," then, was to work at repetitive, simple tasks requiring little discretion.[30] Economic efficiency could be obtained at the cost of making one's work increasingly dull. To be asked to do such repetitive work was often insulting to someone touchy about their social status – for instance, a university-educated gentleman who became an HMI, priding himself on inspecting each school as a qualitatively distinct entity and writing up his reports accordingly. Victorian educationalists saw assessment being partitioned into simpler tasks, each task requiring less judgment. Mechanization, depersonalization, and objectivity reinforced each other.[31]

This is why the spread of written examinations created tension between those hired for their judgment and discretion, like HMIs, and those who sought to promote efficiency by extending the division of mental labour. Such tension is evident in complaints about the "mechanical" nature of examinations, or their "machinery." For instance, in 1867, after returning from a survey of continental schools, Matthew Arnold described the system of payment by examination results by opposing mechanical processes to intelligence:

> In a country where everyone is prone to rely too much on mechanical processes, and too little on intelligence, a change in the Education Department's regulations, which, by making two-thirds of the Government grant depend upon a mechanical examination, inevitably gives a mechanical turn to the school teaching, a mechanical turn to the inspection, is and must be trying to the intellectual life of a school.[32]

Arnold uses the word "mechanical" four times in a sixty-six-word sentence; given that he was also Professor of Poetry at Oxford at the time, it is unlikely he did so carelessly. In his complaints about exams, what Arnold particularly resented was how they restricted his discretion. Such qualities were what made him, and the other Oxford- and Cambridge-trained HMIs, gentlemen. Gentlemen as *gentlemen* were not supposed to be bound by rules and regulations – their trustworthiness was a guarantee of their accountability. Rules and mechanisms were for lower-status clerks.

In a complaint familiar to anyone who grumbles about paperwork and bureaucracy getting in the way of what is considered real work, Arnold called for more "free play for the inspector, and more free play for the teacher."[33] Yet interestingly, where Arnold saw a lack of freedom, Henry Latham at Cambridge saw *more* freedom resulting from delegating authority to an exam, although he too saw an exam as a "relentless piece of mechanism." The exam, instead of the teacher, would compel students (and teachers) to work, which would smooth relationships between students and teachers. Latham even suggested that exams would bring students around to the view that their teachers were allies and guides, not jailors.[34]

We saw earlier how Arnold complained that the Revised Code turned HMIs into "registering clerks, with a mass of minute details to tabulate."[35] On this point Arnold was exactly right. In order to produce usable statistics – details that could be meaningfully tabulated – the discretion of the inspector and teacher had to be curtailed. It is to the tabulation of these details that we turn next.

5 Machining Minds: Commensuration, Tabulation, and Standardization

Of the many features exhibited by every object in a variety of contexts, we abstract one, and consequently, objects qualitatively as diverse as, say, a man's pace, a suit of clothing, a stretch of road, or the height of a tree, acquire a commensurability in our eyes, for we view them from but a single perspective, that of their length.

– Witold Kula, *Measures and Men*, 87

Once an exam was set upon paper, it became possible to "secure uniformity" for examinations in far greater numbers than before. This drive to standardize is the focus of this chapter. Standardization started with basic attempts at first – comparing a set of candidates' answers against what was supposed to be the correct answer, for instance. Then efforts at standardization became more comprehensive, shown by efforts to ensure that exam scores were roughly the same over different years.

The move to standardize exams can be analysed in four stages. The word "stages" is not used to imply teleology or even chronology. Rather, it denotes possibility – the first stages are necessary for there to be second stages, second stages for third stages, and so on. First are the creation of *standards*, or the notion that examinees ought to know certain things or possess skills to a certain level. Standards were made desirable not simply by making their attainment worth something; they were also given moral weight. Common standards made possible the second stage, *commensuration* – a decision to assess otherwise diverse candidates by a quality held in common. Such an act is described in the chapter's epigraph from the historian Witold Kula.[1] Schools tend to group students by age, implying this factor is the most important quality that students possess in common, and more important than the many differences between each individual student. By setting out topics of assessment, achievement exams were assessing these topics as commensurable qualities. For instance, when Philippa Fawcett scored above the senior wrangler on the 1890 Mathematical Tripos, this work was held to be more important than her gender, which was why her victory mattered to so many people. Commensuration made possible *standardization,* our third stage – a word that denotes the actual work carried out to make comparisons between commensurate groups seem reasonable and trustworthy. The grade – or, rather, *a* grade, since there can always be other possible standards against which people can be assessed – had to be made level. And finally, standardization made possible a fourth stage, *quantification,* which is the comparison of commensurable qualities relative to one another at extremely fine gradations. Again, for the quantitative measurement of minute differences between candidates, not only did there have to be a standard being assessed against – this standard had to be seen as more important than candidates' other qualities.

What seems to have resulted from this change was the increasing mechanization of the conception of talent and ability; that is, as the steps to assess knowledge and talent became rationalized and machine-like, participants began to assume that what was being assessed was somehow mechanical. This point supports two other directions of scholarship: William Ashworth has suggested that early nineteenth century mathematical reforms aimed to show that the human mind could be subject to the same kind of scrutiny that could be afforded to a factory or scientific object.[2] (As we saw in the previous chapter, one of these reformers, Charles Babbage, also sought to create machines that "thought," or at least performed elementary mental operations such as arithmetic.) Meanwhile, Lorraine Daston and Peter Galison have suggested that by the mid-nineteenth century many thinkers aspiring to objectivity also aspired to act more and more like machines (the achievements of the examinees writing the Mathematical Tripos, for instance, were celebrated as feats not merely of intellect, but also of machine-like physical endurance).[3] The definition of just what constituted talent and knowledge seems to have narrowed and the diversity of other forms to have diminished at the same time as standards became more rigid.

Standards

Victorian discussions about schooling were obsessed with *standards.* Were they high enough? Too high and exclusionary? Going up or going down? Even the Revised Code of 1862 described the different levels of attainment as "Standards."

Exams compared students against a given standard, but these examinations came in two very different forms: the pass, and the competitive. In a "pass examination," the standard was fixed to a certain amount of skill or knowledge about a topic. Preliminary assessments administered by universities and qualifying exams set by professions were usually measured against pass standards. When the Society of Arts wished to ascertain if a working-class adult was able to undergo a written exam, the candidate had to take a preliminary exam that involved writing from dictation and performing basic arithmetic; if they met that standard, they moved to the next set of more comprehensive exams. By contrast, a competitive standard moved relative to the performance of all candidates. How one did on the exam depended on how all the other candidates did. The most difficult and prestigious examinations, such as the Mathematical Tripos, were always against a competitive standard. Ralph Lingen likened a pass exam to an assessment of which horses could be used in a cavalry regiment; a competitive exam he likened to a race of those horses.[4]

Reformers such as Edwin Chadwick saw pass standards as inferior to competitive standards, partly because pass standards had to be easier. Thus pass standards could not be relied upon to detect the hidden moral qualities, such as persistence, that exams were also supposed to discover. Most importantly, since candidates knew what the pass standards were in advance, they could tailor their preparation accordingly. Competitive standards, by contrast, were superior because of their inherent mystery: no one knew how well one's fellow candidates would place. This indeterminacy meant that each person had no choice but to prepare more intensely. The disciplined study required to do well on such an exam justified the view of Victorian educationalists that competitive exams indirectly assessed moral qualities; Benjamin Jowett suggested that exams against competitive standards unlocked candidates' industry.[5]

A final reason why competitive standards were seen as superior is that the competition itself over time drove up candidate attainment – that is, competition lifted standards automatically, like an invisible hand. We have already seen that Maths Tripos competition drove mathematical attainments upward at Cambridge, making it one of the world's leading centres in mathematics.[6] There are other, more mundane examples of this process: the University of London's Matriculation Examination (partly a pass, partly a competitive standard) in 1838 required candidates to answer questions only out of the first book of Euclid; by 1862 it required them to answer questions out of the first four books.[7] At Woolwich, a guidebook noted that while in the 1850s it was relatively easy for a well-prepared candidate to get into Woolwich, by 1884 it warned that standards had risen to the point that admission was now possible only for those "who received a really good education."[8]

Commensurability and Categorization

Because Woolwich trained "scientific soldiers" such as artillery officers and Royal Engineers, mathematical competence was important, but there was grumbling about the need to pass mathematics exams to enter such places as Sandhurst, which trained army officers. How could mathematical knowledge actually assist in the field? Such complaints continue today, for instance in modern-day concerns about the apparent irrelevance of many university subjects and degrees – what does, say, a history degree offer to someone looking for a job afterward? To ask this question is to compare degrees against an implicit standard of employability – an important quality, although not the only one.

In 1854 Richard William Jelf, principal of King's College, London, opposed the civil service examinations precisely *because* he saw them as trying to square studies and pursuits "which are incommensurable, to establish a common standard of value."[9] Such complaints about common standards of value bring us to the topic of commensuration – a choice to compare against a specific standard, be it Euclid's Proposition Six, knowledge of dogfish anatomy, or future earning potential. Commensuration requires an act of abstraction: one's attention is focused on a specific quality, while at the same time ignoring all of the *other* manifold qualities of a person or an object. Commensuration therefore entails a "harsh editing" of information and contexts now deemed extraneous in that new scheme.[10] For instance, the modern tendency to compare degrees by their potential future earnings means one will also ignore each degree-holder's ability to, say, organize a protest, win a card game, or compose a sonnet. Or, to use a different example, Karl Marx held that the unit of commensuration was the amount of labour going into the creation of something. Labour, in this view, was an abstraction: a common property that turned items into "commodities." Presuming labour as a common property worked to diminish all the other diverse properties of these items – making this new class of things called "commodities" seem more homogeneous than it might otherwise be, with the only relevant difference becoming how much labour went into the creation of each thing.[11]

The epigraph by Kula gives four entities – shirt; a section of road; an aspen tree; a person – that are difficult not to see as anything *but* heterogeneous and qualitatively distinct. Yet Kula points out that it is possible to compare them using a feature they all hold in common, such as length. Length can be expressed as an abstraction (a number expressed in a given standardized system like metric) and then measured.[12] Still more decisions must be made, of course, choosing not only the feature but *how* the feature is to be measured – is a person's "length" assessed in terms of her height or her waist size? What about a shirt or a section of road? Kula's point can also be applied to the commensuration of the knowledge or skills held by people: one can choose to compare a specific skill or realm of knowledge against a given standard, be it knowing Euclid's Proposition Six or dogfish anatomy.

Despite the language of "choice" used above, obviously very few people ever get to choose what is deemed commensurate. Usually one simply acquiesces to what is considered commensurate as part of a very long institutional and cultural history. If a standard has been used before for commensurability, then it's often convenient to use that standard rather than something more suitable – and the longer a standard has been used for commensurability, the more difficult it is to discard it. This was demonstrated by the near-universal insistence that students know the first four books of Euclid's *Elements* for numerous examinations, despite that work being loathed not only by most candidates but also by most mathematics teachers.[13] What made the *Elements* a de facto "common standard of reference," despite its "defects and difficulties," was its sheer longevity, according to Isaac Todhunter, the Cambridge mathematician and examiner, although one might expect him to say this, given that he was an editor of an extremely popular version of the *Elements*.[14]

Despite the topic's seeming lack of relevance, aspiring cavalry officers, the sons of Trinidadian planters, and civil servants had to know Proposition Six or Caesar's *Commentaries* because such subjects were already widely used as standards with which to compare different people. Such a paradox had been recognized by other proponents of exams: in defending the use of Classics for East India Company exams, Thomas Macaulay argued that had Cherokee, rather than Greek, been better known in the British Isles, then candidates would have been tested in Cherokee.[15] This paradox helps explain the longevity of numerous standards that people would gladly be rid of: inefficient or irrational conventions often live on because of their past usefulness. Many examiners, teachers, and candidates had come to coordinate their own decisions and tasks against the Classics or Euclid standard. Any move to a more suitable standard would thus require much collective effort. In other words, such standards had become *entrenched*.[16]

The more widely known a subject, the more people could be compared against a standard of knowledge in that subject. This was how exam systems came to be seen as having the capability to "discover" people who had previously been "hidden" – members of groups that had been formerly kept apart were now "back-bracketed" when compared against a newly important form of knowledge.[17] To test against a standard of knowledge of Euclid was to both elevate the importance of Euclid and diminish the importance of manifold other qualities, whether it be height, social class, hair colour, or drawing ability. Expressed more generically, "talent" was supposed to be a quality discoverable by exams: advocates for the civil service exams spoke in 1853 of their ability to discover the "latent talent of the country."[18] As he propagandized for the new Society of Arts exams, James Booth likened them to mineral surveys, capable of uncovering talent from previously hidden depths. He gave the case of how the exam discovered a bookseller's shop boy so skilled at mathematics that he was hired as an assistant observer for Kew Observatory; similarly, the Cambridge Local exams were praised for discovering young men of mathematical "genius" who had never intended to go to university.[19]

Struggles over commensuration underlay various discussions about just what and who were eligible for examination. While some wanted exams to compare only acts and not people, others conflated the commensuration of *persons* with the commensuration of *performances*. Thus examiners were faced with the problem of whom to admit to examinations (what about women?), while "reformers" pushed to have heretofore unrecognized people (such as women) admitted so that only their acts would be tested. This battle over commensurating *persons* versus *performances* will dominate Chapter 9.

It seems that judgments about which people were legitimately commensurable were shaped by the ideological spectrum running from "conservative" to "liberal." Conservatives tended to conflate the commensuration of persons and the commensuration of performances. Liberals tended to separate the two. Hence more conservative schools such as Oxford saw certain groups of people as incommensurable – not formally admitting anyone who did not subscribe to the thirty-nine articles of Anglicanism. Slightly more liberal schools such as Cambridge might allow non-Anglicans to register (like J.J. Sylvester), but they could not obtain a degree. Non-sectarian institutions such as the University of London saw far more types of people as commensurable, thus making it more "liberal." For instance, in 1845 the senate of the University of London allowed four "native Hindoo students" from the Medical College of Calcutta to complete their medical education at University College, accepting their knowledge of Sanskrit as the equivalent of Latin. Although two of them dropped out, the other two, Bhalanoth Bose[20] and Gopal Chunder Seal, both passed in the first class.[21]

For the University of London, with some slight allowances, men like Bose and Seal were deemed commensurable; such "liberal" institutions saw race and ethnicity as qualities overshadowed by other qualities such as medical knowledge. The students' sponsor vouched for their "good reputation among their countrymen," an allusion to another quality that males were supposed to have at the time: existing or latent gentlemanliness. Indeed, possessing the polite quality of gentlemanliness was one signifier of commensurability: in other words, only gentlemen ought to be tested against one another. Such a belief was strongly (but privately) held by Trollope; he rejected civil service exams for all because "the system of competitive examination supposes there is no difference" between gentlemen and everyone else.[22] He saw the two groups as qualitatively different and thus incommensurable.

What about gender and commensurability? Trollope considered the prospect of examining women as so outrageous that it became a plot point in his *Three Clerks*: Mr. Jobbles [Jowett] falls out with Sir Gregory Hardlines [Trevelyan] by suggesting that "female competitors might, at some future time, be made subject to his all-measuring rule and compass."[23] To be sure, many women saw themselves as perfectly able to participate in such comparison. Although 1850s Oxford and Cambridge were clearly unwilling to accept them, the liberality of the University of London, given its charter statement that all "distinctions of mere class and denomination are abolished, and those of locality modified," was promising. Yet London's willingness to include males such as Bose and Seal contrasts with its reluctance to include women in this mix. In July 1856, two years before the appearance of *The Three Clerks*, Jessie Meriton White sought to be examined and become a candidate for a diploma in medicine at the University of London.

> Sir – Can a woman become a candidate for a Diploma in Medicine, if, on presenting herself for examination she shall produce all the requisite certificates of character, capacity and study from one of the Institutions recognised by the London University? An answer will oblige/ Yours respectfully,/ Jessie Meriton White[24]

White's letter was forwarded to the university lawyer, who referred to the university's charter's call to "hold forth to all classes and denominations of Our faithful subjects, without any distinction whatsoever, an encouragement for pursuing a regular and liberal course of

Education."[25] The charter should be interpreted, the lawyer believed, in light of convention and "ordinary meaning," and so the phrase "without any distinction whatsoever" did not apply to *women*. They could only be admissible with some sort of explicit declaration and debate.[26] White did not press this decision further, and when Elizabeth Garrett Anderson applied in 1868, appealing to the charter, William Benjamin Carpenter, the registrar, formally referred back to White's earlier attempt.[27]

Thus although the University of London liberally allowed the comparison of males from different classes, denominations, and ethnicities, women were deemed incommensurable. The same reasons were given ten years later, in 1868, as to why women should be denied admission for degrees – indeed, even prohibited from writing the Matriculation exam (since regulations stated that any candidate who passed it was entitled to become a student). No one opposed the drive to improve women's schooling; rather, it was claimed that "it was inexpedient to enunciate the principle that the education of men and women must go in the same rut," and that the "mental gymnastics for the male and female sex should be as different as the bodily."[28] Because of the incommensurability of women – linked to the purported sacredness of their role in society – they were deemed ineligible to be examined alongside men.

Espeland and Stevens note that the claim of incommensurability is often used to create boundaries and forbid meaningful comparisons, and shrewdly point out that as commensuration is extended into certain realms, there often emerges a reaction to put other things off-limits as incommensurable. They give the example of market exchangeability – one would not sell a child, a park, or a sacred museum piece, because they are deemed "priceless."[29] A similar case seems to have occurred with Victorian exams: the prospect of extending examinations to female candidates provoked a moral panic over the apparent physical harm that examinations would inflict on female constitutions. Numerous people – often doctors, invariably male – claimed that women's health would be harmed by writing examinations. Despite being empirically refuted, this concern persisted because many held onto the intuition that the constitutions of men and women were qualitatively different and therefore incommensurable.[30] Chapter 9 shows how the push to demonstrate that women were able to do just as well as men on elite mathematics exams was part of a larger reforming campaign to show that women were worthy of commensuration with males.

Standardization

Not to be confused with standards, the push for *standardization* was a key European achievement of the later nineteenth century. Jon Agar has called this move the "invention of methods of exporting similarity" and considers it a hugely important yet overlooked achievement of the period. Specifically in the case of the sciences, historian Thomas Kuhn considered forms of standardization made possible by tabulation and inspection to have created a "second scientific revolution."[31] Yet in the case of examinations, we have seen how the sectarian and disunified state of schooling in the British Isles prevented any central agency from directing curriculum or assessment. The prospect of such an agency horrified many educationists, with the militaristic uniformity of French schooling often held up as something to avoid.

Instead, examinations – or, more precisely, the potential rewards to be earned by taking exams – were an alternative route to standardizing what was learned across the British Isles. In Latham's words, "The examination is the engine, not the fuel" – that function was taken by prizes. Many candidates admitted that prizes kept them studying, with Charles Keetley's student guide confiding to its readers that a £5 prize on an exam written in youth might eventually be worth an extra £1,000 a year in income.[32] Publicizing the names of prizewinners, as well as the rules and required materials for the exams, would indirectly pull candidates towards higher attainments; as early as 1847 Booth noted that such publicity for exams would "induce a conformity to their rules and a sameness in the courses of study pursued by candidates."[33]

To assure candidates and observers that exams were worth taking, not only was there publicity but an aspiration that candidates take their exams in conditions made as uniform as possible. Extensive rules, for instance, governed how exams were to be run, limiting the discretion of any one person. Early rules for the running of examinations, such as those published in 1858 for the Society of Arts, were comprehensive. They specified the seating of candidates (by candidate number, sitting at least two feet apart, no two students from the same school could sit together); who was to provide the material used by each candidate to furnish answers (for each exam "paper" worked, the organizing board provided three sheets of foolscap and one sheet each of scribbling and blotting paper, while candidates brought their own pen and inkstand); how the answer-sheets, upon being written on by candidates, were to be returned to the examiners (paginated, grouped by subject, along with a candidate declaration sewn to the bundle with a green silk twist and needle); the routines that the organizing board had to follow in order to guarantee public trust in the examinations (after an exam, each board member had to sign a declaration attesting that all rules had been followed).[34]

Comprehensive and strictly followed rules were intended to assuage concerns that without an examiner present, confidence in the exams would fall (in the case of exams run by prestigious institutions such as Oxford) or never emerge (in the case of less-prestigious groups like the Society of Arts). For the Society of Arts, the promise that its "competition will be perfectly fair" meant well-publicized cancellations of exams at the slightest hint of an irregularity. An error made at the Oldham Lyceum in the geometry and algebra papers of 1858 meant the cancellation of the papers of candidates 405 (Thomas Crellin) and 406 (Ralph Crompton). The examiners emphasized that the cancellation did not result from anything the candidates did; rather, it was a misinterpreted rule, and to assuage their feelings, the Society gave Crellin and Crompton the money they *would* have received had their papers not been cancelled.[35] With such measures the Society sought to assure the public that every passed exam, regardless of where it was taken, would attain as close as possible a "uniformity of value." To this end, over the years the rules of the Society of Arts' exams became ever more comprehensive: more detailed eligibility conditions; suggestions for how *exactly* organizing boards should be constituted; how "previous" (i.e., preliminary, or qualifying) exams should be run. Its rules even began to specify the desired forms of handwriting: bold, without loops, long tails, or flourishes.[36]

Standardization also occurred when one exam system imitated the rules of other exam systems. The exams of the University of London were explicitly modelled on Cambridge's, and carried out by numerous Cambridge graduates who had become London examiners. Specific rules governing the exams of Woolwich Military Academy, where Royal Engineers were trained, were adopted by DSA exams, which were then enforced by those same Royal Engineers who gone through Woolwich. The organization of the DSA exams owed much to the workings of the Oxford Local exams, but then the architects of those Oxford Locals had, after all, copied the organizational plans not only of the exams of the Society of Arts and the College of Preceptors, but also the earlier ones of the Education Department. This copying of rules occurred in well-established exam systems, as we see in 1880 when the University of London's registrar asked for and received a copy of Woolwich's rulebook governing its exams.[37]

The rules governing the materials used in examinations further assisted with standardization. When in 1867 the first overseas

University of London BA exam was held in Mauritius, London's registrar sent blank registration sheets and admission cards, as well as a template that a Mauritius printer could use to run off blank answer booklets. One year later this same template was sent to the ten Canadian centres running London's Matriculation exam.[38] Practical examinations were far more difficult to standardize – perfect uniformity was sometimes unattainable. For instance many botany exams have candidates identify different plants, meaning numerous similar specimens must be available. Yet global examinations made the uniform provision of such specimens difficult. What if a plant used in London was unobtainable in Colombo or Kingston? To compensate, examiners sent out lists of plants that could be found in these locations, requesting that organizers obtain them in advance. When new London exam centres were added, such as in Cape Town or Barbados, a local botanist would be recruited, and London would send him a list of examinable specimens. He would then modify the list to fit the ones actually found there.[39] Another case saw attempts to standardize laboratory-based practical exams after someone realized that candidates being tested at the same laboratory where they trained had an unfair advantage over others. They would be more familiar with those instruments. To counter this, examiners shuffled the instruments around the laboratory.[40]

Standardization was also difficult when it came to the relative difficulty of different subjects, or perception thereof. On the various papers of the University of London's Matriculation exam, failure on a single paper meant failure on the entire exam, a problem that prompted internal discussion on how questions in different subjects could be made of similar difficulty. Later, in 1882, after Thomas Thorpe and James Dewar were censured by the London senate for putting unexpected questions on their organic chemistry exam, all exam questions had to be submitted for scrutiny to an examination subcommittee and London's registrar before being printed. At London's examiners' committee meetings, members from different subjects gave friendly tips to one another on the best ways to draw up questions.[41]

Ever more meetings were created by the need to standardize marking. The DSA meetings have been recounted above. At the University of London, where questions were co-authored by two examiners, explicit marking procedures were to be followed. Candidates usually clearly passed or failed the paper, and so the opinion of only one examiner was required; if the result was unclear, the examiner sent the paper to his counterpart. If they disagreed, a third examiner was brought in. London's honours papers and the exams of high-scoring candidates were to be independently assessed by both examiners.[42] How would a new examiner ensure that his marking corresponded to that of more experienced colleagues? The examiner and exam coach Philip Magnus suggested that the group of examiners randomly select several answers to the same question, give each a numerical mark, and keep this sample nearby as a ready-to-hand reference.[43]

Quantification

Magnus's suggestion brings us to the topic of quantification: in order for exam results to be seen as meaningful, there has to be a belief that the properties being quantified have some degree of uniformity. The University of London's Matriculation exam was marked in a simple and standardized way: each hour was worth 100 marks. So the relative value of the subject was revealed not only by the numerical value of the paper but by the length of time allotted to answer it. Natural philosophy took three hours and was worth 300 marks; so too for chemistry, English grammar, and geometry. Latin grammar was less valuable, at two hours and 200 marks (but when combined with Greek translation and grammar, the combined subjects in Classics were worth 700 marks). The 1867 Matriculation exam took twenty-eight hours to write and was thus worth a maximum of 2,800 marks, although the top score ever attained was 2,300, and only nine candidates had ever received over 2,000 marks.[44] Meanwhile the Society of Arts marked each paper out of 100, with 75 or higher being a first class, and below 30 a failure.[45]

There is some dispute over where the use of numbers to assess exam answers first emerged – likely in Paris, although possibly Cambridge – but provenance is less important than understanding what numerical marks could be used to do. They could be aggregated to yield a single score for the entire exam, a single number that served as an abstraction of performance. Commensurability is implied by the use of integers: the ranked series in which they appear creates the context that gives that single number meaning (the meaning of an exam score of "9" will differ radically depending on whether that exam is out of 10 or 100). The use of a single number to abstract overall exam performance was done with the Cambridge Locals: totals for all of a candidate's papers in each subject were aggregated into a single number, then used to rank them. Yet numbers are not always necessary for ranking: Oxford's internal examinations used Greek letters to grade and even rank exam performances. Ranking was possible because Greek letters also belong to a ranked series: an alpha denotes a higher place than a gamma by virtue of its appearing earlier in the series.[46] However, there was no commonly accepted way to aggregate multiple Greek letters to generate a single symbol – an abstraction of a more complicated performance. It is far easier to do this with numbers. Even so, there has to be a conscious decision to allow separate numerical scores to be aggregated and yield a single score, as Cambridge did. Their rivals at the Oxford Locals did use numbers to assess candidate performance, but Oxford resisted combining those scores into a single number. Instead there were four division lists – one per subject – and a candidate's rank was given in each list. Oxford examiners were doubtless referring to the rival Cambridge system when they sniffed that their system was better than "a number of things amalgamated in a lump."[47]

Numbers denoted not only an exam subject's value but also the supposed value of that subject to society. When it devised the Indian Civil Service exams, Macaulay's committee was accused of using the value of each subject as "an index of the relative value to be attached to each separate department of knowledge."[48] This was not denied. For this exam, then, which by 1860 was worth a total of 7,375 marks, knowing mathematics (pure and mixed) was worth 16.9 per cent while Sanskrit (language and literature) was worth 6.8 per cent.[49] Yet the apparent precision of these numbers hid the fact that these different numerical values emerged out of squabbles between examiners over different subjects: the most visible outcomes of negotiations between factions. The point is analogous to historian Ken Alder's observation that the relative prices of the ingredients of a loaf of bread, or the sizes of gauges and jigs in the manufacturing workplace, also emerged out of past conflicts and negotiations, and the power of each faction relative to each other – farmers versus millers in the case of bread, and shop-floor workers versus managers in the case of gauges and jigs.[50] Indeed, if we extend this argument, it seems possible to chart the rise and fall of a subject in cultural importance relative to other subjects by looking at changes in numerical marks. What was worth knowing yielded relatively more marks. The rise in cultural importance of the natural sciences was reflected in the rising numerical value of those papers on the Indian Civil Service exam.[51] Conversely, the declining cultural importance of Classics relative to the natural sciences was shown by the decreasing numerical value of Classics questions.

A further advantage of using numbers to denote value in an exam, as opposed to other symbols such as Greek letters, is that when aggregated they also allow averages to be calculated. Such averages can in turn be used in a temporal series. Was a particular standard of attainment rising or falling? Rising or falling numbers would tell the story.

In 1867, for instance, London's registrar reported that the rejection rate on the Matriculation exam was in a fairly stable range of between 33 and 40 per cent each year. He also noted that there had been academic improvement because the standard to gain a first class on the BA exam had risen from 1,000 out of 2,400 marks in 1853 to 1,100 in 1867.[52]

However, such apparent numerical precision was not the whole story. Examiners did not always feel bound to strictly follow a numerical scheme: this would mean they were "mechanically" following it, after all. Despite the exactitude of London's Matriculation being worth 100 marks per hour of work, its registrar reported that the number of marks required to pass – "about" 1,000 out of 2,800 marks – was only an informal guideline. There were numerous cases in which a candidate "scraped through" in each of his papers and passed; he would fail the entire exam only if one examiner said "we must reject him."[53] Another way in which numbers disguised messier processes was in their allocation: did one add marks together to create a total, or subtract marks from an ideal answer? When it came to national schools, the HMI Joshua Fitch recommended that examiners add or subtract depending on whether the ignorance they detected was benign (a lack of knowledge or an inability to articulate oneself) or malignant (pretension or "demonstrative ignorance of one who thinks he knows"). Only in the case of malignant ignorance should marks be deducted.[54]

Examiners wished to be seen as individual agents with discretion, or – to use a more context-appropriate term – as gentlemen.[55] This meant they often freely ignored whatever marking system was being used. For the Oxford Locals, the numerical marks were often reweighted after the exams had been written, in post hoc recognition of difficult questions; meanwhile, at the Cambridge Locals, an examiner was "expected to represent his estimate of the candidate in marks" and to continue the Cambridge tradition of giving over 100 per cent for an exceptional answer to a mathematics question. Yet an examiner was able to assign a value to an exceptional answer only after seeing some of the other answers.[56] One amusing case of refusing to use exam results and marks "mechanically" was shown by Henry Latham, who was himself an examiner for many years. In conversation with another Cambridge examiner, George Forrest Browne, Latham declared that while examiners were all very good at yielding lists of marks, he knew a man better than they did and so did not let exam results entirely drive who received fellowships to Trinity Hall.[57]

One way to preserve this cherished discretion was to keep the numerical value of questions secret. The marking scheme of the Cambridge Locals, for instance, was kept private, with the note "For the Use of

Examiners Only ***It is particularly requested that this list may not be shewn to any one."[58] When the Oxford Local examiners testifying before a Parliamentary Select Committee were repeatedly asked to disclose their specific weighting coefficients, they always refused.[59] Making such formulas public would not only tie their hands but make it easier for examinees and their teachers to game the system.

Another way to preserve examiner discretion was to refuse to return marked answers after an exam. In May 1860 the Conservative MP for Portsmouth rose in the House of Commons and asked for the release of a few copies of successful civil service exams. One of his constituents had failed the exam three times and believed he could make it on a fourth try if he had better models for his answers. The MP noted that he had been forced to initiate a motion in Parliament because his letters to the Civil Service Commissioners had gone unanswered. The motion was opposed by Liberal MPs such as Gladstone and Lowe, who pointed out that the since the commissioners' role was judicial, they required the independence and discretion that publicity would strip away. The motion was defeated fifty to eighty.[60] A similar case occurred twenty-four years later at the University of London: a candidate failed the chemistry paper of the 1884 Matriculation exam, despite teaching chemistry elsewhere and holding a diploma from the Pharmaceutical Society. So he requested that his paper be assessed by another examiner, which the university refused. The candidate visited the registrar's office at least three times and even threatened lawsuits. He was ignored. In a final plaintive note, he asked why the university so strongly opposed the scrutiny of completed exams by outside experts.[61] It seems that the university did not respond, but we can answer the question: it is because such external scrutiny would have meant the loss of power by the examiners, mechanizing their role.

Yet even just the publicity of overall scores and candidates' rankings could lead to greater pressure for a uniformity that reduced examiners' discretion. In 1882 a group of "Concerned Schoolmasters" compared the results of London's Matriculation exam over several years, and discovered that its pass rate fluctuated by as much as 17 per cent a year.[62] In their "Letter of Concern over Unevenly Marked Examinations," the petitioners noted the unlikelihood that the intellect and knowledge of the average exam candidate would fluctuate that much from year to year. They shrewdly pointed out that such uneven marking was likely caused by a "fluctuation of the vague standard floating in the minds of its examiners." They proposed that the university set "certain narrow limits" based on past results, which would reduce the variation and thereby "secure even-handed justice."[63] Thus the push for greater uniformity and more restrictions on examiners' discretion was prompted by demands for fairness.

Such petitions and concerns must have been on the mind of F.Y. Edgeworth when, in his 1888 and 1890 papers "The Statistics of Examinations" and "The Uncertainty of Examinations," he made the same point as the petitioners: the law of large numbers meant that the fluctuations were probably caused by the examiners, and these would diminish if more examiners marked each paper. He wished to apply error theory from the physical sciences, suggesting it might be possible to discover a coefficient of minimum error where one could say a candidate was justly or unjustly marked: a state of affairs where one might imagine if, say, a different staff of equally competent examiners had marked the exam, a different mark might have been obtained.[64] We have come a long way from the viva voce medical examination discussed in the previous chapter. Edgeworth's intervention is noted in John Roach's comprehensive history of examinations in England.[65] But one must not overlook what made Edgeworth's proposal possible in the first place: all the work done to ensure all candidates were being tested under the same infrastructural conditions.

Tabulated Poetry

Meanwhile, the reduction of examiner discretion made possible the comparison of different schools on commensurable qualities such as exam passes. These statistics could be used to rank schools and candidates that had not been previously compared. Just what could be done with a more "mechanical" system of examinations was shown by none other than one of Matthew Arnold's less-famous brothers, Edward Arnold, who was himself an HMI in Devon, Cornwall, and the Scilly Isles. In the very same 1867 report in which Matthew Arnold criticized an overly "mechanical" system, Edward also made a report on the 192 schools he visited. But his description differed substantially from his brother's. Edward did not simply present tables of numbers, but used the numbers to compile averages to represent the performance of various Plymouth-area schools on certain categories. This made it possible for him to show, for instance, that when it came to "accuracy of instruction" in arithmetic, Plymouth-area students were below the average performance of children at all national and Church of England schools (a 60.5 per cent pass rate in Plymouth against a 73.8 per cent national figure). He could also argue that there had been a decline of arithmetic scores between the 1865–6 and 1866–7 school years of 5.9 per cent (from 66.4 per cent passing to 60.5 per cent).[66] As

Schools arranged according to

Per-centage of Passes on Examination.	Per-centage of Children grouped in Standard IV.—VI.	Average of combined Per-centages in Columns 1 and 2.
1. St. Neot's, boys - 96·9	Mawnan - - - 62·5	Dartmoor Prison, boys. 70·
2. Gulval - - - 93·3	Plymouth, St. Andrew's Chapel, girls. 61·7	{ Morval - - - 69·8
3. St. Agnes - - - 92·4	Plymouth, Grey, boys 60·3	{ Gulworthy - - 69·8
4. Morval - - - 92·2	Dartmoor Prison, boys. 53·1	Plymouth, Grey, boys 68·8
5. St. Neot's, girls - 91·1	St. Mary-Tavy, girls 51·8	Holne - - - 64·5
6. West Alvington - 90·6	Tavistock, boys - 50·5	Marazion - - - 63·7
7. Holne - - - 89·8	Gulworthy - - 50·	Gulval - - - 63·3
8. Gulworthy - - 89·6	Penzance, boys - 49·5	St. Mary-Tavy, boys - 62·45
9. Lamerton - - - 89·4	Tavistock, girls - 48·	Plymouth, Charles, boys. 62·15
10. St. Ives, boys - - 89·	Morval - - - 47·4	Penzance, boys - - 62·05
11. Stratton - - - 87·4	Mabo - - - 47·	{ Scilly Isles, Tresco - 61·
12. { Torpoint, boys - 87·3	Boscastle - - - 46·2	{ Probus - - - 61·
13. { Scilly Isles, Tresco - 87·3	Crowan - - - 44·8	Grade - - - 60·95
14. Probus - - - 87·2	Marazion, boys - 43·6	Plymouth, St. Andrew's Chapel, girls. 60·5
15. Carnmenellis - - 87·1	St. Mary-Tavy, boys 43·3	Tavistock, boys - 59·85
16. Dartmoor Prison, boys 86·9	Salcombe, girls - 42·	Devoran - - - 59·3
17. Bude - - - 86·8	Stonehouse, girls - 41·1	{ Mabe - - - 59·1
18. Bideford, boys - - 86·8	Scilly, St. Mary's, boys. 40·8	{ St. Stephen's - - 59·1

Figure 3. "Schools arranged according to per-centage," from Edward Arnold, "General Report, for the Year 1867, on the Church of England Schools Inspected in the Counties of Cornwall and Devon," in *Report of the Committee of the Privy Council on Education* (London: Eyre and Spottiswoode, 1868), 48.

shown in Figure 3, the newly meaningful tabulated information even allowed Edward to rank each school performance by its aggregated examination results: by "per-centage" of exam passes at each school (ostensibly a measure of "accuracy in instruction") or by percentage of students in the three upper standards (classes), ostensibly measuring "extent and range of instruction."

The point here is not that Edward Arnold was particularly original in his use of averages to show differences – rather, where many historians have quoted the more famous Arnold for his criticisms of the constraints imposed by an increasingly "mechanical" schooling system, a far less well-known Arnold at the same time was showing how useful numbers could be produced as a result of such a system. Edward did not use the numbers unthinkingly, and repeatedly warned readers that the numbers were imperfect. But he did argue that the table could give a "fair estimate of the relative merits" of his district's schools. Indeed, one later historian judges that Edward was far more aware of a problem that his brother Matthew did not seem recognize: that although many of the best students were learning to read, write, and do arithmetic, a large proportion of other children in those same schools were learning hardly anything at all. For all of its terrible faults, the Revised Code was doing something it was supposed to – pushing teachers to also focus on the quiet children who were easy to overlook.[67]

Some people saw standardized tests as imposed from outside, thus mechanically restricting the discretion of educationists and students; others saw them as creating new ways to meaningfully measure and improve human performance. Standardization could both oppress and emancipate, sometimes at the same time. It is in these opposed reactions to examinations becoming more uniform, more mechanical, that the "Janus face" of standardization is revealed.[68]

6 Thin Descriptions: Credentials and Other Signals

There never was a bishop who gave his son the best living in the diocese who did not conscientiously believe the son he gave it to the best and fittest man for the post. The difficulty is to get other people to believe it.

– Walter Wren, "Education and Examination," 401

"They cannot know what a nice boy he is or how much thought and sense he has. They have only the examination to go by." Thus lamented Anne Wynne about the trials faced by her son Edward. It was 9 July 1856, and "Eddy" had just finished the entrance exam for Woolwich Military Academy. It had been Eddy's second try, and he was certain he had failed again. Although he had practised drawing machines, he felt he had been misled by the syllabus: machine drawing was worth fewer marks this year, and landscape drawing was not optional, as previously declared, but mandatory. Eddy and his mother saw this change as unfair.

Parents, and anyone witnessing a child's development over a long time, have good reason to feel they know them better than most – their quirks, their long-held dreams, their secret strengths. Such a belief underlay Anne's point about the poverty of knowing someone only through an exam. Yet how much of this is sentiment and wishful thinking? Anne was aware of these tendencies among parents and pre-emptively conceded the point – she confessed that every other parent of Woolwich candidates probably also wholeheartedly believed their own son to be well suited for the Academy. So how good a judge could she be, really?

Anne's letters were addressed to the natural philosopher and examiner John Tyndall. He happened to be an examiner on this same Woolwich exam, setting and marking the questions on its natural philosophy paper. Tyndall and the entire Wynne family had been long acquainted: by 1856 Anne's husband, George, was himself a lieutenant colonel in the Royal Engineers, and he was Tyndall's earliest patron. When starting in 1840 Tyndall worked on Royal Engineer-directed surveys of England and Ireland, he requested testimonials from George; in the early 1850s he continued to seek career advice from him. In turn Tyndall did what he could for the Wynnes: when he became Professor of Natural Philosophy at the Royal Institution of Great Britain, he helped George on technical questions such as metallurgy, and he hosted Eddy and his brother when they visited London.[1]

In our more suspicious age we may wonder if Tyndall interceded on Eddy's behalf. Of this there is no sign. Anne's letters never request anything that would constitute a breach of trust or other form of misconduct. She simply asked Tyndall if he could provide notice in advance of the formal publication of results whether Eddy had passed or failed. And to be sure, Tyndall did help Eddy in ways that were unlikely for other candidates – he provided hints on which fields Eddy might improve, and told Anne to send Eddy to Queenwood to get private mathematics lessons from his friend Thomas Hirst.[2]

Eddy did fail this second attempt, but on the next he succeeded, entering Woolwich to become a Royal Engineer. After all, if he wished to become one, passing the exam was the only route available. Such entrance examinations were becoming an obligatory point of passage to enter an institution or demonstrate one's competence.[3] For Eddy's was but one case in the British Isles in which exams started to become the preferred – even the only – way in which to demonstrate the qualities necessary to be admitted to schools like Woolwich, to be hired in certain workforces like the civil service, or to gain a degree testifying that one was competent in a certain subject. Exams were clear tests against increasingly explicit standards, be they in double-entry bookkeeping or elementary inorganic chemistry. Another point that this chapter discusses is the chasm that divided how people came to be understood. Anne's letter to Tyndall is a complaint about that chasm. On the one hand, exams seemed to discover and provide clear and explicit signs of knowledge or character; on the other hand, a person could be intimately assessed by another over the course of a long acquaintance. Which form of knowledge is more valuable to society? Anne realized that tests such as the Woolwich entrance exam rewarded the external, the explicit, and the formal at the expense of the internal, the implicit, and the informal.

Hence where Chapter 4 depicted Victorian exams as rituals upon which performances could be fixed upon paper – as photographs of the mind – and where Chapter 5 looked at some ways in which mass exams were standardized, this chapter explores how examinations became the favoured way to ascribe competence and other forms of knowledge to a person. Exams became rituals of certification, and trust was placed in the resulting pieces of paper. It is a truism dating back to at least Max Weber that in the nineteenth century, paper credentials became increasingly important in a more populous, urbanized, and anonymous world.[4] But underlying the relative value of each credential was a larger system that formed the credential's reference group. Some certificates and degrees were more trustworthy than others. By the 1850s such verification work was becoming ever more difficult as ever more candidates were able to travel longer distances.

One response to this problem was the 1858 Third Charter of the University of London, also known at the time as the "New" Charter. By making the passing of its exams the only necessary condition to earn a BA from the University of London, the Third Charter is said to have made the University "the birthplace of long distance learning."[5] But the charter was controversial because – as many of its opponents pointed out – it did not so much create a system of *education* as it created a system with which to verify the ostensible *results* of education. It was not distance education, but certification-by-examination: an assay. Not only did the Third Charter make it possible to earn a BA without ever setting foot in a classroom; it became possible to do so without ever talking in person with a teacher or a fellow student. This impersonal approach to education poses a striking contrast with the intimate knowledge of a person gained by a parent or familiar friend. Unlike Anne, however, the Third Charter's proponents saw this sole reliance upon exams as a benefit.

Saving the Labour of Verification

The taxi driver can rarely identify *where* the University of London is and the foreigner can never understand what it is. So relates one historian of the university, Negley Harte, who refers to it by its evolutionary stages: "Mark I," between 1826 and 1835, and "Mark II," between 1836 and 1900.[6] The university during Mark I had various names that we lump for simplicity's sake into the moniker "London University." It emerged out of a series of meetings in 1825 and 1826, where Henry Brougham – polymathic founder of the *Edinburgh Review* and reformist Whig MP – plotted with representatives from the Jewish, Catholic, and nonconformist communities to create a secular institution. Since none were Anglican, none could matriculate at Oxford or gain a Cambridge degree. London University opened its doors in Bloomsbury in October 1828, charged with being "godless," the "synagogue of Satan," and "stinkomalee."[7] The hurlers of such invective welcomed the creation of the staunchly Anglican King's College the next year; given that its patrons included the Duke of Wellington, three archbishops, and seven bishops, it received a Royal Charter almost immediately. But no such document was given to its London University rival. After the 1830 election, wrangling at the cabinet level between Brougham (now Chancellor) and his opponents held up any similar document. Five years later, they finally negotiated Mark II: London University became (London) University College, whose students, along with King's College's, would be examined by scholars who would "perform all the functions of the examiners in the Senate House of Cambridge."[8] Indeed the model of Cambridge was always in the background, with ten of the original thirty-eight founders being graduates of its best-known college; Harte calls the University of London's Mark II "Trinity College, Cambridge, writ metropolitan."[9] Thus was constituted the University of London's board of examiners, whose operations comprised the bulk of the new institution. The University of London Mark II was an examining board, testing "externally" what was taught "internally" at its two member colleges. The university's emergence as an examining board was a clever compromise to the problem of sectarianism, allowing students to study in an environment befitting their cultural and religious preferences, while nonetheless setting a uniform and formal way to assess students. This "federal" mode of organization was adopted by other new universities throughout the empire, including the Queen's University in Ireland in 1850, the University of Toronto in 1853, and the University of Madras (Chennai), University of Calcutta (Kolkata), and University of Bombay (Mumbai) in 1857.[10]

The University of London's structure and function was formalized by its First Charter of 1836, which declared that the university's purpose was to ascertain, "by means of examinations, the persons who have acquired proficiency in Literature, Science, and Art." Thus someone intending to enrol at University College or King's College would take the Matriculation exam; upon passing it, and after presenting a certificate of attendance from one of those two schools attesting that he had completed the required courses, the student became eligible for further university exams. Passing those exams led to a degree in arts, law, or medicine from the university. Another sentence of this charter preserved the secular intent of the original 1825 projectors of London University: to "hold forth to all classes and denominations of Our faithful subjects, without any distinction whatsoever, an encouragement for pursuing a regular and liberal course of Education."[11]

As we have already seen, despite the charter's inclusive phrase "without any distinction whatsoever," university candidates consisted almost entirely of Jewish, Catholic, and nonconformist males.[12] Over the next forty years came challenges emphasizing the gap between the formal language of the charter and the conventional practices that belied it. These challenges always focused upon exams and certification rather than classroom attendance. Recall how the language of legal formality pervaded Jessie Meriton White's 1856 letter: could she become the first female candidate for London's degree in medicine if upon "presenting herself for examination she shall produce all the requisite certificates of character"?[13] The charter was invoked to allow Bose and Seal to enrol after their eligibility was questioned.[14] In another case, the charter's claim of inclusivity was taken very seriously regarding students with disabilities: when the blind Daniel Conolly petitioned to write London's Matriculation exam, a committee was struck to devise a system to conduct the exam that would "place him as nearly as possible on the footing of ordinary Candidates."[15]

A commitment to greater access to formal schooling and the formalism and explicitness of written exams was not rooted so much in respect for rights or diversity – instead, it was intended to promote "free trade in education."[16] The "free traders" behind the University of London sought to align the university with market forces such as supply and demand, prices, and competition. Doing so would help overthrow the existing educational "monopolies" of the universities of Oxford and Cambridge. Examination results made free trade in education possible by explicitly showing what students had learned. One free trader was Robert Lowe, who eventually became MP for the University of London. We have also seen how Henry Cole and Lyon Playfair did much to develop payment by results at the Department of Science and Art, helped by their ally Edwin Chadwick. But a key ally of Lowe, Cole, Playfair, and Chadwick was George Grote, the banker, Radical politician, and historian of ancient Greece. Eventually he became London's vice chancellor. In the 1830s Grote frequently joined Cole, Chadwick, and John Stuart Mill on country rambles and evening talks; around 1838, Playfair joined these men at a London dining club. Calling themselves "friends in council," they met to discuss political economy.[17] Grote is the central figure of this chapter because it was he who did the most to change the University of London's charter in 1858.

The charter was changed to solve a problem then plaguing the university: the untrustworthiness of many of its incoming students' certificates. London's charter of 1836 allowed only students enrolled at University College and King's College to take the University of London's exams and thereby gain its degrees. But that charter suggested that other colleges might "affiliate" with the university, giving their own students the right to take University of London exams and acquire its degrees as well.[18] The process was simple – the school sent a request to the university senate, describing its courses and how they were taught. This request was then voted on by the senate. Between 1836 and 1856, numerous colleges from across the British Isles and then the empire became affiliates. After attaining this status, an affiliate's student who wished to write one of the university exams went to its buildings at Somerset House. Upon arrival, he would present two forms to the registrar: a certificate in which two respectable people attested to his moral character, and a certificate from his college stating that he had been a student there and had fulfilled its requirements.[19] The rule was closely followed: in 1839 one student who came all the way from Dublin to write the Matriculation exam was turned away by London's registrar, R.W. Rothman, because he had forgotten the necessary forms.[20]

Yet this strict adherence to rules was paired with a lack of curiosity about precisely *what* these pieces of paper actually signified. When in 1841 James Booth, as principal of Bristol College, sought the college's affiliation with the University of London, he asked Rothman what London defined as a legitimate student, and how Bristol should record student attendance. Rothman replied that these matters were Bristol College's own business: the University of London recognized "as students those persons who produce Certificates from the Colleges that they are or have been Students; and has nowhere defined the meaning of that term."[21] Rothman loaned Booth a completed certificate of attendance from University College upon which to model Bristol's. Booth asked no further questions, and Bristol College duly became an affiliate.[22]

This preference for form over content continued as new candidates showed up with their certificates to be examined. The numbers of candidates gradually grew with the increase of affiliated colleges, a process driven by the expansion of the railway network (Figure 4).[23]

In January 1846, Montreal's McGill College became London's first overseas affiliate.[24] By the early 1850s about four to five new affiliated colleges were added each year, about one at each senate meeting. No application ever seems to have been rejected. By 1857, colleges affiliated with the University of London included not only University College and King's College but also the universities of Oxford, Cambridge, Durham, and Dublin; various Queen's Colleges (e.g., Belfast, Galway, and Liverpool); Stonyhurst College in Lancashire; Owen's College in Manchester; and the Working Men's College in London. While most were seen as reputable, some recent additions had unsavoury reputations. One new affiliate, the Bishop Stortford Proprietary School, was

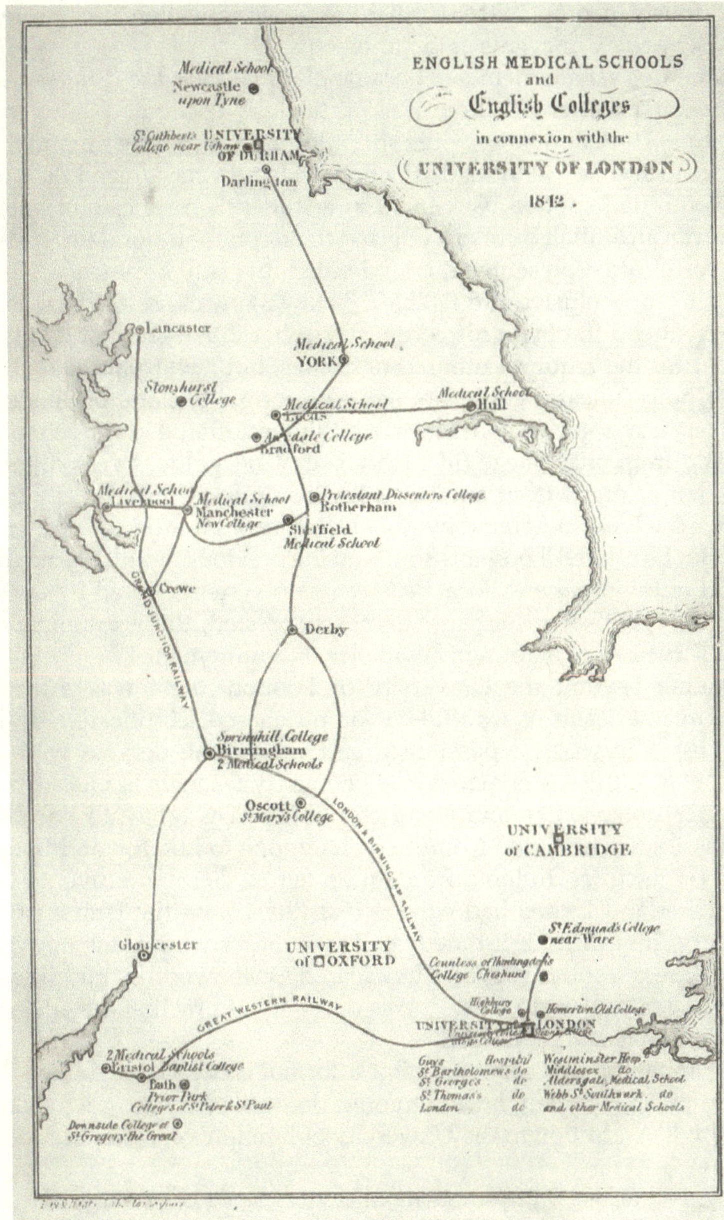

Figure 4. "English Medical Schools and English Colleges in Connexion with the University of London, 1842," from V.A. Huber, *The English Universities*, vol. 2 (1843), 564.

"notoriously incompetent to furnish academic education," according to one 1857 review of events at the university.[25]

Numerous senate members became alarmed over the differences in affiliates' reputations.[26] Grote was particularly concerned: how could one assess student quality while still fulfilling the charter's mandate to educate everyone "without distinction"? He now recognized the problem Booth had spotted back in 1841: a student's certificate of attendance from an affiliated college referred to no common standard. Instead each certificate represented an "undefined diversity of practice," making each "inconclusive and fallible."[27] How many classes did a student have to attend to obtain his certificate? What if that student had been present for the required number of classes, but never paid any attention? The certificates could not say. The formal declaration made by each certificate starkly contrasted with the intuition that the schooling received from one college (like University College) was very different from that received from another (like Bishop Stortford). In July these problems were enumerated by Robert Barnes, a London medical graduate. But he offered a possible solution: the previous month, Barnes had helped out with a new "local" examination system offered by Oxford University in Exeter. Perhaps, Barnes suggested, these examinations could serve as a new form of "evidence of acquirement."[28]

Looking beyond the University of London, there was a further problem: the issue of fraudulent, or purchased, certificates. By the early 1850s this was a possibility, and even some degrees were for sale. In 1853, the mathematics and chemistry teachers at Queenwood College, Thomas Hirst and Heinrich Debus, were aghast to learn that its Classics teacher, John Mummery, had gone to London and bought for £25 a doctoral diploma from an unnamed German university. As both Hirst and Debus had earned real PhDs from the University of Marburg, they were infuriated: such purchases would devalue their own degrees in others' eyes. They feared Queenwood would be seen as a place of "sham doctors."[29] Was diversity of practice so wide as to include purchased certificates?

The apparent gap between what a formal credential signified, and what a person actually knew or could do, was one reason why Grote proposed an addition to the University of London's charter: Clause 36.

> 36. We do further will and ordain, that persons not educated in any of the Institutions connected with the said University shall be admitted as Candidates for Matriculation, and for any of the Degrees hereby authorised to be conferred by the said University of London other than Medical Degrees ...[30]

"*Not* educated in" was the key phrase of the addition. Anyone not registered at an affiliated college would be allowed to write London's Matriculation exam. If they went on to pass further exams, they could earn its arts or law degrees.[31] This meant that someone could earn a London degree in arts or law without ever setting foot in a classroom. By making the only condition of earning a London BA the passing of its exams, Clause 36 standardized the process by which it was earned. More accurately, to follow language used by Lawrence Busch,[32] Clause 36 *simplified* university schooling – eliminating the problem of "diversity of practice" by only requiring evidence of attainment. For a certificate to be trusted, it had to be clearly shown as the outcome of a respected test.

Clause 36 was officially opposed by both University College and King's College. A memorandum from the Council of University College complained that this clause surrendered to "Free Trade in education"; it meant that the senate had come to look upon the university "as a machine constructed solely for prescribing and conducting Examinations." The council's use of the word "machine" reveals a concern about a mindless and repetitive process. Reading alone was not enough to acquire a BA – what about meeting and talking with people who thought differently, or knew more than oneself? Such contact gave the student a chance to imitate the habits of conduct and thought that learned people displayed, especially the ability to see a subject from another person's perspective. The benefit of such contact was not simply intellectual – it was also a moral benefit.[33] Exams could assess only intellectual attainment. John Henry Newman's views about an ideal university are given context when situated against what he and others loathed: the formalism inherent in the push for free trade in education and in Clause 36. Anthony Froude complained that the University of London represented the victory of shallow knowledge – that which was easily examinable, "technically committed to the memory, without thought, without appropriation ... not even the knowledge of what knowledge is."[34]

However, people like Froude and Newman weren't members of the senate. Despite the protests, Grote's side won the senate vote, and the Third Charter was adopted on 22 July 1857.[35] Just over a year later the university followed the example of the Oxford Local exams, allowing its Matriculation exam to be written at satellite centres outside London. The first two were in Manchester and Liverpool, where candidates wrote at precisely the same times as those in London; by 1865 the Matriculation was being written overseas, in Mauritius. The combination of Clause 36 and these new satellite exam centres set in place the framework for what became known as the University of London's "external system." In five year intervals between 1855 and 1870, the number of Matriculation candidates grew from 209 to 428, 616, and 845[36] – rapid growth similar to that of other mass exam systems of the period.

At the time, other institutions tried to simplify their testing as well. Since at least the 1830s medical reformers had complained of the "multiplicity of institutions" – at least nineteen, ranging from the University of London to the Archbishop of Canterbury – that could certify someone as a medic in the British Isles. This diversity strained the public credibility of the entire medical profession. The simplest solution, various reformers argued, was a single general exam to be written by all aspiring apothecaries, surgeons, and physicians.[37] Although such a sweeping simplification never occurred, the 1858 Medical Registration Act led to the publication of names of people passing their medical exams and to a gradual reduction of certifying institutions, once members of the new General Medical Council started visiting the exams of different licensing bodies. The simplification resulting from the Medical Registration Act inspired a similar call for the certification of teachers: in 1862 the College of Preceptors explicitly drew on that Act as a model for how teachers should be certified.[38]

Fallacies of Indiscriminate Laudation

We can move from the descriptive to the analytical portion of this chapter by asking how competence, knowledge, and character were ascribed to a teacher; a surgeon; a student. To someone lacking familiarity with a person, what forms of evidence could convince them that he or she had the qualities necessary to carry out the tasks of the job? Then, as now, there loomed challenges of trustworthy credentials.

One option was the testimonial letter, in which a person of some reputation or responsibility spoke of various qualities in the recipient. In attesting to the qualities of the letter holder, the letter writer attempted to confer trust in him or her. Sometimes such letters were private; for more prestigious and competitive positions, these letters were often circulated and published. We saw how John Tyndall sought testimonial letters from George Wynne as soon as he left home in 1840; by 1851, in his unsuccessful quest to gain the professorship of natural philosophy at the University of Toronto, Tyndall had collected and published glowing testimonials from various European scientific luminaries, including Edmond Becquerel, James Joule, and Gustav Magnus.[39]

Sceptics denied the value of such testimonials as evidence of knowledge or good character. Benthamites led the criticism. They saw testimonial letters as given out too freely, at little personal cost to the writer. Robert Lowe described the testimonial as a "splendid instance of what Bentham calls the fallacy of indiscriminate laudation."[40] Two members of John Stuart Mill's rambling group discussed above felt the same way about testimonial letters; upon taking over the School of Art in 1852, Cole instituted exams after observing that many holders of glowing testimonial letters were unable to draw.[41] His friend Edwin Chadwick saw testimonial letters as "generally worthless" and untrustworthy, not because they contained lies, but because they all depended upon the diverse and idiosyncratic standards of the letter writer rather than on an accepted, single public standard. Hence what one person might consider neat handwriting, another might deem unacceptably sloppy.[42]

In 1856 an author in the *Journal of the Society of Arts*, probably James Booth, declared that every reader knew that "the system of granting appointments on the faith of testimonials is a bad one."[43] Nine years earlier Booth had already called for certificates derived from exams to replace testimonials,[44] and now he proposed that the results of the Society's exams be entered into a public registry. His intention was to give potential employers "precise and accurate information ... as to the attainments and intellectual ability" of certificated candidates.[45] Who might use this? A statement signed by several hundred people – including businessmen, bankers, MPs, the Archbishop of Canterbury, engineers such as Robert Stephenson, and savants like John Herschel and Charles Babbage – promised to use this registry when looking to hire new staff, and would treat Society of Arts exam certificates as "testimonials worthy of credit."[46] The exams established certification-by-examination as a more reliable way to ascertain if someone could actually do what they claimed to be able to do.

The gap between claimed ability, knowledge, and character and what was actually there was analogous, Edwin Chadwick argued, to the problem of adulteration of food and medicine. How could a sceptical Victorian trust that the ingredients were what they purported to be? In 1832 Babbage's *Economy of Machinery and Manufactures* took up this problem. It was widely known that staples such as flour often had ingredients like chalk added by unscrupulous millers trying to save money on grain. To prevent this, buyers spent time inspecting flour for unwanted additives. Yet this task was relatively simple compared with having to verify the purity of more complex substances, such as medicines: a job difficult even for medics. Babbage concluded that government agencies should produce the items themselves: the work of such trusted groups would "[save] the labour of verification" that would otherwise have to be spent by consumers. The entire economy would benefit from the time saved.[47]

Edwin Chadwick explicitly extended Babbage's principle to exams: he thought they saved the labour of verification when it came to people. In 1854 he recalled that during his time as a civil servant in charge of numerous workers, he witnessed enormous differences of talent, judgment, and industry among them. Some were far more effective than others at doing the work asked of them. But Chadwick noted that to ascertain just *who* was effective was complicated and time-consuming, for some good workers were content to labour in "obscurity," while some inept workers actively pretended to be more productive than they were, performing what Chadwick called "eye service." Chadwick considered exams to be a more effective way to ascertain competence, which was why he so strongly supported the proposal to examine prospective civil servants as outlined in the Northcote-Trevelyan report.[48]

Another recurring image used to describe exams, certificates, and the gap between claimed and actual ability, knowledge, or character was *currency*. Single testimonial letters were drawn up according to one person's knowledge of another person, and often did not refer to a shared and agreed-upon value – but exams did. The reference to a shared value made the credentials produced by exams analogous to currency. Variants of this theme likened exams to the stamps ("hallmarks") placed upon metals to denote verified currency. In 1858 the new Oxford Associate-in-Arts designation resulting from its new Local exams was described as "a new coin … struck from a die like [the] old ones, but of a different metal," with the exams "putting yearly a stamp on a number of young men."[49] Robert Lowe described the University of London as a "great intellectual mint, to which gold may be brought from every quarter, and from which, when stamped, it may go current all over the world."[50] The biologist George Rolleston described Oxford exams as "imposing Mint Marks and Trade Marks on men."[51]

As an image conveying how qualities could be reliably ascribed to people, the metaphor of currency was also deployed in calls to rein in needless complexity. In 1839 the *Lancet* argued that having nineteen different authorities certifying medical practitioners was as ridiculous as having nineteen different groups testing gold and silver coins against their own individual standards.[52] Exams were a form of assaying: by preventing counterfeits they reduced deception of self and of others. Stafford Henry Northcote in 1859 saw exams as preventing someone from mistaking "his highly-burnished copper for fine gold."[53] Or exams could discover unrealized riches: when candidates underestimated their abilities it was as though "pure gold passes for base metal," claimed James Booth. As with Australia before the gold rush, "genius, energy and talent … lie buried in the masses around us," but these treasures, "until found and brought to light," were useless.[54] Widespread exams, then, mined for hidden talent and better aligned outward appearance with inner qualities.

What about the gap between what a school promised or advertised and what actually occurred there? This disparity was often decried as "puffing,"[55] and fuelled a wider mistrust of teachers at mid-century. There were some attempts at assessment – from 1836 on, government-funded schools were visited by inspectors from the Education Department as a condition of that funding. But as we saw in Chapter 2, a key criticism by the Newcastle Commission in 1862 was that such group assessments overlooked most of the students in a school because the HMI was pressed for time. Inaccurate inspections were exacerbated by teachers strategically manoeuvring their best-performing students into the most visible places – another case of "eye-service."

At any rate, a school receiving no government funds – which in 1860 meant most English schools – did not have to be inspected at all. No parent wants to send their child to a "bad" school. But how could a parent even tell if a school was bad, especially during a time of such diversity of practice? The point was emphasized by Thomas Acland when promoting his Oxford Local examinations, echoing Chadwick about saving the labour of verification. He reprinted a tradesman's letter likening schools to a consumer product: "Those who buy bread, or meat, or clothes for their boys have some tolerable judgment of the article; but, in buying instruction for them, they buy in the dark."[56]

Since parents were not present in the classroom, it was difficult for them to know what actually occurred there. Usually the only information they had was gleaned from the child's reports, school messages, or advertisements. W.E. Forster – MP, political slayer of Lowe, champion of the 1870 Elementary Education Act, and brother-in-law of HMI Matthew Arnold – related the problem:

> A great number of the parents are themselves utterly ignorant, or at any rate, would themselves acknowledge that they are very ignorant, as to what is a good school and who is a good teacher, and that they are liable to be misled by pretentious advertisements, and by a pretentious scheme of teaching; and they are also liable to be misled by their boys and girls, who come home in the evenings, and tell them they have learnt a great deal, when in fact, they have learnt very little – and the innocent parent believes them.[57]

Meanwhile, John Robson, Secretary of the College of Preceptors, readily agreed when a commissioner summarized his testimony at the Taunton

Commission: "Parents cannot safely be trusted to choose schools and schoolmasters for themselves."[58]

Adding to this mistrust was the problem of educational "quacks," claimed Forster.[59] Just as reformers of allopathic medicine gave heterodox or unlicensed medical doctors this label, so too it was applied to unscrupulous teachers by those trying to reform schooling and teaching. This label, with its implication of widespread fraudulence, was used by everyone from Booth to critics of the private tutoring industry; from Acland's tradesman correspondent to John Stuart Mill.[60] Deceitful practitioners harmed the students they taught and threatened public trust in schools as a whole. To combat "quacks," both school and medical reformers called for a combination of simplification, exams, certification, and publicity.

Examiners as Anti-Anthropologists

Exams, then, were not simply assaying tools with which to distinguish between a single person's genuine knowledge and its appearance. Sets of exams, and their results, could be used to distinguish between different groups of people. The results of exams purported to distinguish between good and bad schools by yielding abstractions of performances. The numerical scores that represent the relative merits of exam answers could be aggregated, from simple counts of individual scores ("ten of the fifteen students passed the exam") to sums of those scores ("the class score was 78 out of 150"). Such sums could be turned into averages ("the mean score was 73 per cent on the Euclid examination") as abstractions of performance of larger groups, such as the Form or entire school. And those numbers furnished representations of even larger group performances, such as school districts.

These obvious points are belaboured to emphasize that Victorian exams belong not only to the history of education, but also to the history of statistics. Consider Alain Desrosières's definition of statistics: the reduction of diverse and abundant situations to a summarized description of them that can be used as a basis for action.[61] Examinations, particularly as they became more standardized and formalized, collected meaningful information that educators could analyse. The abstracted results that exams yielded were seen as more comprehensive than any alternative.

One alternative to exams was the inspection. Although championed by HMIs such as Matthew Arnold, other educationists saw inspections as less effective. Frances Buss, headmistress of the North London Collegiate School for Girls, believed inspections could never replace exams:

"Inspectors could not examine every paper, of every class, in every subject, in such a large school as this, and give marks ascending to a scale." Nor could an inspection accurately assess the school as a whole. (Buss also thought that male inspectors, *because* they were male, would never be able to fully understand schools for girls.)[62] Only the results of trustworthy tests could yield disinterested representations of complicated situations like the "state of a school." In short, inspections alone could not produce usable statistics.

For his part, Robert Lowe viewed exams as instruments that elicited from school managers "confessions which I should have thought nothing but the rack would have extorted."[63] Political decision-makers like Lowe saw exams as generating information that could show whether schools receiving government funds were truly "efficient" at working to explicit standards such as those of the Revised Code. This sentiment continued to be felt long after Lowe had moved on from his Education portfolio: nineteen years after Lowe's 1862 remark, the Vice President of the Education Department, A.J. Mundella, declared that the "number of ignorant and neglected children brought in" to the schools (revealed in 1870 once elementary education was extended still further) was finally diminishing, and "the number of backward children grew less and less." As evidence he pointed to the increase in the percentage of exam passes from 80.4 per cent in 1880 to 81.2 per cent in 1881.[64] Despite such a small increase of 0.8 per cent, Mundella regarded the numbers as reliable indicators of a large social change, or at least sufficient evidence to warrant a public pronouncement of improvement. Such sentiments exemplify Theodore Porter's view that schools have been fashioned into places where exams, and the ostensibly objective numbers generated by exams, are meaningful.[65] Although Porter's comment is directed at contemporary schools, the point also usefully illustrates the use of exams in the mid-Victorian British Isles. In short, schools – as institutions using exams – were not simply formal places of education. They were becoming places where reliable statistics could be collected.

Might we follow the point of James Scott's *Seeing like a State* and depict the push for exams as part of a larger move by the British state to make performances "legible" – that is, to shape and direct society on the basis of abstract representations?[66] To be sure it is governments that are in the strongest position to generate and use statistics, being one of a tiny number of institutions with the longevity and funding to count and classify people over the long term, as well as holding a monopoly on the legal violence necessary to compel anyone resisting this drive to count and classify.[67] Yet it was not merely the government that pushed for such descriptions. To see exams as inflicted "from above" by a dirigiste state is insufficient: as we have seen, statistics and certificates generated by exams were used by parents, teachers, administrators, and even students to make choices and guide their actions. Indeed, the whole point of using exams to standardize schooling was to avoid a central government ministry prescribing curricula, as in France. In 1877 one English headmaster explained to a German visitor what exams had done: "We were before like the Israelites when they had no king; every one acted as he thought fit: since the net of examinations has been spread over our schools we know what we have to do."[68]

What, then, should we call the credentials, the marks, the letter grades, the percentages that resulted from a person's performance on an exam and that are seen as standing for a person's knowledge and competence? While often – though not always – numerical, they are all abstractions, summaries, simplifications. To call them "signals" is useful, and in this light we can see exam results and their resulting credentials as "thin descriptions," to adopt another of Porter's notions – an ironic counterpoint to Clifford Geertz's famous "thick description." As is well known, Geertz coined the phrase to denote the understanding of a person's action in terms of its meaningfulness to that person. Thick description relies upon the idea that culture is a context within which social events, symbolic actions, behaviours, and institutions can be oriented and thus intelligibly described.[69] By contrast, thin description gives outsiders the ability to choose and act upon information without needing a deep understanding of the immediate setting.[70] Thus Robert Lowe or Henry Cole, despite not knowing the specific conditions under which schooling occurred, could know the general state of those schools – or *believe* that they knew the general state of schools. Thin description is necessary for public knowledge of, or democratic control over, some complex realm.

The tension between thick and thin description is captured in Anne Wynne's complaint to Tyndall that the Woolwich examiners could not know Eddy as a complex and unique individual. For what knowledge could be "thicker" than a mother's assessment of her own child? All the Woolwich authorities had to go by was the exam, which decontextualized, which flattened, which thinned, all of Eddy's knowledge into a simplified report of a set of performances that had been conducted under highly restrictive and stressful conditions. Indeed, if one is to extend Porter's take on Geertz, Victorian examiners seem to have worked as *anti-anthropologists*;[71] that is, they constructed systems that allowed the commensuration of diverse members of multiple groups into a single larger class, "flattening" descriptions of them to more effectively compare individual members of that new class. If we are to take Marilyn Strathern's definition of ethnography as "open-ended" inquiry that allows for the unpredictable and contingent,[72] then perhaps Victorian examiners are more precisely characterized as "anti-ethnographers." Such thinning and flattening seems to obviate the need for any deeper, localized knowledge about the context of those individual members. It is an impoverished understanding. But on the other hand such thinning and flattening seems to be the only way to make possible, on a mass scale, descriptions that can be taken as trustworthy and used to make decisions by non-experts outside that system.

Public Opinion, Reactivity, and Looping

Chadwick claimed that exams extended Babbage's principle of "saving the labour of verification" from consumer goods to people. Yet the situations are not identical: the verification of people is not the same as that of flour and medicine. Unlike those inanimate substances, a person is an intentional agent who can change his or her behaviour on the basis of an assessment.[73] Indeed, a person usually does reflexively respond to being tested against explicit standards.

The point about reflexive responses brings us back to another cautionary statement made by the Council of University College London regarding Clause 36 of the Third Charter. The councillors warned that relying solely upon written exams as evidence of a good education was worrisome because such tests tended to reward the *display* of learning – especially the recounting of facts. Exams were far less effective at demonstrating the "habits of thought" by which such facts were learned. For proponents of Clause 36, the display of knowledge was partly the intent – after all, in 1857 Lyon Playfair claimed that paying for successes on DSA exams would induce teachers to experiment with different ways to teach science, leading to the spread of the most effective pedagogies.[74] Yet we saw how the UCL Council had presciently warned that students would begin to anticipate which questions would appear on an exam, often by studying previous exams in the same subject. In so doing they would tailor their learning to the exam.[75] Students preparing to be tested would begin with the end in mind – passing or triumphing on that exam – and devise their strategies accordingly.

Chadwick's analogy is imprecise in another way. As an inanimate substance, flour cannot warn other kinds of flour about impending tests; medicine cannot communicate with other kinds of medicine. But people can and do. For many proponents of examinations, the ability to communicate the content of exams and their results counted as a point in favour. Such exchanges made it possible for individuals (students, parents, teachers) to compare their behaviour with others, thereby prompting them to change their behaviour where they saw fit. For exam proponents, this tendency for examination results to loop back and change behaviour was yet another reason to favour them. For it meant that publicized exam results could be allied with the force of "public opinion," thereby reforming institutions. Exams became engines, not cameras.

Because of the role of public opinion, one source of exams' power was the mass media. Exam proponents inspired by Ricardian political economy claimed that publicizing exam results would stimulate competition between individuals and between schools, and that such competition would lead to better schools. Clear and public exam results would be seen and understood by parents, who over time would choose higher-performing schools. A simulated market would thus automatically improve schools: a belief articulated by both Booth, in 1847, and Lowe, in 1868.[76] Such a belief is still held by present-day believers in school choice, or by those who publish rankings of schools' performances on standardized tests.

There were numerous cases of results widely publicized with the intention of stimulating further competition. My favourite example appeared on the front page of *The Times* on 5 August 1858, shown in Figure 1: a juxtaposition of competitions. The first two columns list the ranks of candidates who completed the second Oxford Local Examinations; the third column, the places of the horses who raced at the Brighton Plate. One can thus compare the seventh place finisher of the senior division – S.H. Stocks of Huddersfield Proprietary School – with the seventh place finisher of the one-mile race (handicap 50 sovereigns), Mr. King's "Naughty Boy."

Another way that exam results were exposed to public opinion was the directory. First appearing in 1861, *Crockford's Scholastic Directory* was, in the words of one reviewer, a "new idea" revealing a "new-born want of advancing civilization": where should middle-class parents send their child to school?[77] The directory answered this question by clearly identifying which schools were better and which ones were worse – that is, better or worse when compared against explicit standards tested by exams. Advertisements for schools had existed for a very long time, but sceptics derided those ads as "puffing." By contrast, directories used exams as common tests of school "efficiency." Another directory's author argued that only schools that taught students properly could get them to pass such ordeals as the Local exams.[78] While the modern-day equivalent of such exams, standardized tests, are often depicted as alien impositions upon teachers by distant administrators at the behest of ignorant politicians, in the mid-nineteenth century "external" exams were not always seen this way. Some teachers saw good exam results as clear demonstrations of their teaching abilities, and thus as evidence to substantiate their schools' advertisements. In a number of cases, when parents of an especially promising student were reluctant to pay the expensive fees for a Local Examination, the teacher paid out of his or her own pocket.[79]

With the Locals and other exams testing against explicit standards of what constituted a "good education," and with directories publishing these results, tabulated displays of exam results began to appear by the mid-1860s. In 1864 one educational journal printed a ranking of teachers based upon results of their students on the Cambridge Local exams. To recognize different levels of difficulty, the table's compiler had allocated points according to the exam – a first-class honours on the senior exam yielded eight points; a pass on the junior, only one. This particular table is noteworthy for listing teachers and tutors not only across the British Isles – it also included candidates from Trinidad (coached by Horace Deighton), where the Cambridge Locals were also being written. Thus it had become possible to make ostensibly precise comparisons of teaching ability across oceans. Again, this is not a case of a government simply imposing "legibility" upon its subjects in the manner of *Seeing like a State*: the person who compiled this table was not a government agent, or a Cambridge don, but a hobbyist vicar who found the results interesting enough to send to the *Educational Times*.[80] By 1876 we see more guidebooks displaying such tabulated results; F.S. de Carteret-Bisson's *Our Schools and Colleges* displays a large table of various schools' results on exams.[81] The principle of success or failure – according to these exams' explicit standards – was clearly shown.

Exam results placed students, teachers, and schools into a larger, reactive, community. Some members of this community were eager to participate in their own commensuration and altered their behaviour accordingly. As the standards and tabulations became more important, a teacher or school manager – whether they wished it or not – had to take those into account if they wished to become successful in what was rapidly becoming a simulated marketplace. Indeed by 1867 one directory stated that it would list no new school *unless* it had proved its "capacity to satisfy recognized examiners."[82] The more seriously exam results were taken, the more publicity they received, which made the exam more important ... which tended to attract more people who took the exam results seriously. Over time such reactivity entrenched exams, first as important routines with which to assess performance, then as necessary ones. The same logic is encountered in modern-day rankings of law schools or global universities, which so often compare apples to oranges or make other dubious assumptions. Even if administrators do not wish to participate, rightly doubting the validity of such rankings or worrying about what the rankings will do to their schools, they are often forced to do so if they wish to keep their jobs.[83]

As students and schools shaped their own definition of what a good education meant against these tests of explicit standards, so too was there a change in the definition of a good student or a proper school. Roles and institutions came to be redefined: successful learning and efficient teaching could be identified by exam successes. Exams came to dominate schooling because, as intentional agents, people began with the end in mind – of performing well on exams.

Indeed, in keeping with the notion of "free trade in education," it was only rational for candidates – now starting to see themselves as economic agents – to consciously devise strategies to pass the exams, and to do so as efficiently as possible. They began to show "looping" behaviour, focusing their energies on displaying the laudable qualities that exams were supposed to assess. Part III therefore shifts our perspective towards the various reactions that Victorian examinees, such as Eddy Wynne, had at being so extensively tested.

PART THREE

Examinees

7 Learning and Earning: Coaching, Cramming, and Arms Races

[The Hon. Mr. O'Donnell, MP] was of [the] opinion that the whole system of competitive examinations required re-casting. It was something like the contest between big guns and iron plates; the guns had got ahead of the plates, and the crammer had got ahead of the competitive examination systems.

– 264 *Parliamentary Debates* H.C. (3rd ser.) (8 August 1881), 1297

Ten years after Eddy Wynne wrote the entrance exams to Woolwich Military Academy, the pseudonymous candidate "Cheth" took those same exams and passed. He reported to his old school magazine how he had prepared. His parents paid 120 guineas to send him for a year to the Reverend Adolphus Smith's Establishment for Young Gentlemen. Each year, about 100 candidates attended to learn how to pass the entrance exams for various military academies. Where Eddy Wynne was retaught mathematics by a future Fellow of the Royal Society, Thomas Hirst, Cheth could call only upon the Reverend Smith and his long-suffering assistant, Patrick O'Brien. Cheth imagined that O'Brien saw Smith's Establishment "as near an approach to purgatory as could be imagined,"[1] and recounted O'Brien's various stratagems to get his unwilling charges to learn the required material. Cheth did his part to make the Establishment purgatorial for O'Brien. When he locked himself in his room to escape O'Brien, the assistant asked questions through the door. When early one winter morning Cheth and a friend tiptoed out of the house to skate on a nearby lake, O'Brien caught up with him and asked him to repeat formulas as he skidded along the ice. When the train carrying the candidates to Chelsea Hospital (the exam location) began to pull away from the station, O'Brien ran alongside the train hurling questions through the open window at Cheth and his fellow candidates.[2] Happily, Cheth did pass the exam.

The only way for Cheth and Eddy Wynne to get into Woolwich was to pass the entrance examination. But there were other incentives to take examinations, such as the prospect of rewards and other prizes: fuel for the engine, as Henry Latham's saying went. At Latham's Cambridge college, Trinity Hall, his fellow tutor, friend, and sometime-rival for its mastership, Henry Fawcett, saw examinations in the same pragmatic way. After all, as the seventh wrangler of 1856 and a political economist, it was self-evident to Fawcett that schools were also governed by economic laws. Like Latham, he thought it absurd to complain that only a few students were purely devoted to their studies: exams had set teachers and students in a race; the prospect of obtaining prizes made knowledge a "marketable commodity."[3] Precisely *what* was learned was not the point: what mattered was whether mastery of a given body of knowledge was widely accepted as a standard with which to gauge a candidate's character and temperament – the reason why, as we have seen, using Euclid's *Elements* to teach geometry still survived. Leslie Stephen, Latham and Fawcett's colleague at Trinity Hall, noted the incomprehensibility of Euclid to the average student, who wondered why he should learn Euclid or "get up Scripture history any more than why he should stand upon his head and repeat abracadabra seventy times running."[4]

Having to take so many exams could even destroy one's love for a subject. The philosopher Bertrand Russell later noted that studying for the Tripos had led him "to think of mathematics as consisting of artful dodges and ingenious devices and as altogether too much like a crossword puzzle."[5] Just after completing his final paper for the 1859 Mathematical Tripos, James Wilson had a breakdown that "entirely swept away my higher mathematics"– and he became senior wrangler that year.[6]

Yet ambitious candidates were forced to perform highly on exams even if they saw them as meaningless. As a result reflexive performance – "looping" – against the formal standards assessed by exams began to emerge. People were endlessly inventive at ways to polish their performance on the formal and explicit attributes of the exam. They might trade tips or practice, or gain additional instruction from coaches. Indeed, even if candidates sincerely believed in an exam or the goals of the exam system, it was necessary to polish their outward, formalized, performance. Some bought into the moral economy of the examination system; others saw it as meaningless. What resulted was a split between performance and content, between the appearance of knowledge and the actual possession of it. Hence just as administrators sought to make a complex system legible through abstraction – thin description – so too did the candidates working within that system try to provide back to those bureaucrats thin descriptions of their own. We might call this "thin *knowledge*," something that formally satisfied the standards being examined. Ironically, such responses undermined the moral qualities that exams were supposed to indirectly measure.

The gap between appearance, form, and performance on the one side, and authenticity and content on the other has existed as long as there have been exams. There are obvious parallels in other realms: for instance, between those who act on their religious belief and those who pay mere lip service to it by reciting catechisms. At any rate, examiners and other educationists were immediately aware of the problem of thin knowledge. They called it "crammed" knowledge, and the people who provided this appearance of knowledge became "crammers."[7] There emerged shadow institutions run by such crammers. Many coaches hotly denied this label, arguing that they were in fact superior teachers. As evidence for this higher pedagogical skill, they pointed to ... their students' exam successes.

As exams grew in popularity, concerns mounted about candidates who only appeared to know something in order to pass. There resulted a moral panic over cramming and crammers. Examiners put in place measures to thwart cramming, wishing to reward only those who authentically possessed knowledge. Examinees and their teachers, in turn, devised countermeasures. An arms race emerged. This chapter is about that arms race. It studies those who came close to deliberately violating the rules that governed examinations, but who did not actually do this: those violations are described in the next chapter, on examination cheating.

Self-Help

We begin with candidates devising in advance the most efficient way to perform to the required standard. One rational option was to do the bare minimum required for a pass, and direct the rest of one's energies to more pleasurable outlets. This had been the warning about mere pass exams by Edwin Chadwick and Henry Moseley, and it was borne out numerous times. Such strategizing occurred even at institutions founded as part of a moral crusade. In Chapter 9 we will see how Cambridge's first unofficial women's college, Girton, was created with the explicit goal that women could, and should, perform to exactly the same standards as men. Yet in 1882 one Girton tutor remarked that some of his students viewed Cambridge's heavily mathematical "Little-Go" exam as injurious; they frequently asked him, what was "the very smallest amount [of mathematics] that will get me through?"

At lessons, a recurring question was whether the current mathematics lesson they were enduring was absolutely necessary to pass the Little-Go.[8] If there were multiple possible exams to take, the rational option was to ascertain which path required the least work for the most reward. In the language of political economy, students sought to know which route "paid" best.

As long as there have been exams, it has been known that key to improving one's score is to practice being examined. HMI Joshua Fitch exhorted teachers to familiarize national school students with being examined by providing model answers to questions and showing the order of reasoning in an answer.[9] Growing familiarity with exams led to higher scores; for instance, in 1858, the first official year of the Oxford Locals, only 280 (37.3 per cent) passed out of 750 candidates. But by 1863, 487 (63.9 per cent) passed out of 762 candidates. The examiners noted that this was not because the exam's standards declined, but because in 1858 the "circumstances were new and strange" for most examinees. For some of them the general notion of a formal exam must have been unfamiliar, and for others the specific expectations and format of the exam were unfamiliar. But in subsequent years the candidates were better prepared by their teachers, had a better idea of what to expect, and deployed more suitable repertoires. Exams became more difficult or asked new questions. To use one examiner's phrase, the candidates, the schools, and the exams "started to fit to each other."[10] The sociological term "tuning" might also be used to describe this process of adjustment of each party to one another, as each individual action comes together into a collective act.[11]

As examinations became more widespread, more publications appeared to assist candidates through the process and to show them what actions to add to their repertoires. Sometimes these works were stand-alone guidebooks, and sometimes they were articles appearing in magazines that emphasized their own flinty-eyed pragmatism. For instance, the *Practical Teacher* (many of whose readers sought University of London degrees to advance their careers) printed many such guides. One popular version of this genre was penned by successful examinees who related specific tips about just how they got high marks on their various exams. A common theme of such guides was their exhortation that it was not enough to merely study: one had to also know how to display what one knew on an exam in formats that would please the eyes and sensibilities of the people marking that exam. As writers, exam-takers were told to consider their audience of markers.

When one sat down at the table, one should *strategize*: for instance, do the easiest questions first. In mathematics exams, that meant doing the "bookwork" (definitions) first, then moving on to the problems, comforted that one has already done half the work. *Remain confident*: competitive exams meant that no one knew what the outcome would be, so it was a mistake to worry. *Make every moment count*: two minutes lost may mean five or six marks lost, which might mean the loss of a rank or two on a candidate list. One should thus write as fast as one could.

But at the same time it was important for a candidate to *be aware of the marker's needs*. Thus one should present answers in formats easy for the examiner to read and understand, perhaps even "in a tabular form" so that the examiner may be made happy "by your evident desire to please him and save his time." Handwriting also mattered: the examiner might have to mark 2,000 or 3,000 papers, and "writing hieroglyphics" would only alienate him. Another way to keep the examiner happy was to structure one's answers: clearly show where a question ended and another began by enumerating the responses. Mathematics answers should be underlined, the steps in reasoning arranged sequentially, and each step given its own line in the answer book. Diagrams were often useful too, in that "a few well-drawn figures" could illustrate understanding better than pages of writing (as long as one didn't forget to clearly label the diagram).[12]

Candidates with initiative often prepared for exams by talking with past successful examinees to gain "tips." Such preparation was a time-honoured practice at Cambridge. One candidate for the 1867 Mathematical Tripos, writing pseudonymously as "Examinee," recounted how, after passing the first, easier portion, he now had to study for the second, more difficult, part. He contacted a friend, a high wrangler in a previous year. The friend told Examinee to focus on sound and gave him some books and notes; he then explained a page or two of formulas, and told Examinee to memorize them and practise writing them out. A good friend indeed, the former wrangler checked Examinee's formulas to ensure their correctness. Such studying filled Examinee's mind with a jumble of mathematical terms and formulas "which I understood about as much as the man in the moon." On the first day of the second part of the Tripos, Examinee saw that one of the questions did indeed call for one of the formulas his friend had given him. Examinee wrote it out, leaving a "seemingly accidental smudge over a questionable 'd' in the middle" to obscure that he had memorized it. On successive days of writing, Examinee found that his friend's tips paid off again and again; on the final paper of the Tripos, Examinee was delighted to see a question on sound to which his friend had supplied the answer, and he wrote it out triumphantly. His high spirits were brought down slightly when, upon handing in his paper as he left the Senate

House, one examiner showed it to another and laughed. Unfortunately, we do not know how "Examinee" ultimately did.[13]

Knowledge as Food, or Bricks?

By obtaining "tips" in advance and memorizing formulas he himself admitted he did not understand, had "Examinee" really learned the mathematics that the Tripos was supposed to test for? On the matter of thin knowledge, was it possible to feign knowledge without understanding it – Victorian versions of the "Chinese room" puzzle?[14] As examinations grew in popularity, such questions began to be asked more often. Thomas Hirst wondered about this as he coached candidates for the Woolwich entrance exam in physics and mathematics. In addition to Eddy Wynne, he coached the son of Lord Francis Gordon. Hirst found that the boy, "naturally sharp enough," had learned the first three books of Euclid by heart, "but I am convinced he *never understood* a proposition."[15] Meanwhile, as his friend John Tyndall marked some of the exams written by Hirst's students, he mused whether candidates for his own exams actually understood the required material. One of his 1859 natural philosophy questions asked,

> Whence does the odour arise which is produced by the combustion of a sulphur match? Describe the characters of the two principal compounds of sulphur and oxygen.

Tyndall wrote out one candidate's answer:

> From the fumes arising from the decomposition of the sulphur – Sulphur contains a large quantity of sulphuric acid, and Brimstone obtained from active volcanoes is one of the chief substances.[16]

The candidate's answer recalls a tactic as old as Molière's dormitive virtues – providing an answer by rewording the question. Chapter 3 also showed similar tactics by examinees.

To many educators such an appearance of knowledge was *cram*; the acquisition of this simulacrum was *cramming*. Being crammed was one predictable way in which candidates prepared to be thinly described by an exam: to a question a candidate provided a short answer, an abstracted version of that knowledge, to demonstrate its possession. In turn the examiner then rated that answer with a numerical representation that was itself a further abstraction. At any rate, various contemporary definitions of cramming can be found. Isaac Todhunter equated cramming with learning by rote, using "memory alone" where "reason ought to have shaped and controlled" the product.[17] Civil Service Commissioners (defending themselves against charges that their own exams promoted it) defined cramming as the acquisition of "a superficial appearance of knowing subjects of which one is ignorant, committing to memory disconnected facts, learning by rote answers to expected questions."[18]

Despite complaints about memorization, learning by memorization was not always seen as problematic. It had been encouraged in Enlightenment studies, informed by psychological theories of associationism that held ideas were linked together in the same order in which they were acquired. Historian Matthew Eddy has shown how Scottish medical students were encouraged to copy out passages from books, because these notes would order how they would later recall that information. The order in which one memorized ordered the growing "stock of knowledge" in one's mind.[19] As written examinations became more popular, however, an ancient metaphor of knowledge also grew in popularity. Knowledge was depicted as a substance being forced into a container-like mind. That substance was *food*.

The metaphor of knowledge as food seems to be at least as old as the Stoic philosopher Epictetus, who in his *Enchiridion* (135 CE) commanded would-be philosophers to show their learning by their actions, not words: sheep did not vomit up grass to show what they had eaten. Instead they transformed it into wool and milk.[20] In 1706 John Locke described humans as intellectual ruminants – while reading gave us the "materials of knowledge ... it is not enough to cram ourselves with a great load of collections; unless we chew them over again, they will not give us strength and nourishment." Both philosophers were quoted by Victorians; Locke's advice was repeated in an 1871 guidebook for University of London examinees, while Epictetus's sheep were cited by Lyon Playfair in 1873 (by now disenchanted with examinations) to celebrate exam candidates who actually understood the material.[21]

The depiction of knowledge as food generated related metaphors.[22] Learning was akin to eating. The omniscient Aristotle, one Victorian noted, would have "borne an enormous gorging." True understanding was fully digesting that learning. The truly educated person was someone whose knowledge was "rightly assimilated," so the goal for schools was to supply "nutriment for the mind."[23] Thus crammed knowledge meant that it remained in one's memory – which functioned like a stomach – in undigested form, and thus lacked meaning. Instead, a crammed candidate had a "mass of undigested facts committed to memory', or 'propositions to be learnt ... eviscerated of all meaning and then, like preserved meats, packed into the smallest possible formula."[24] Such retention was considered harmful. Mr. Minusex, the protagonist's rival in Trollope's *The Three Clerks,* made himself ill and could no longer be examined by Jobbles, for Minusex had "so crammed himself with unknown quantities, that his mind – like a gourmand's stomach – had broke down under the effort." On the day of the test, Minusex remained home, "sobbing out algebraic positions under his counterpane."[25] The crammed candidate even had an iconic representative animal – the bird stuffed with food. Sometimes the Strasbourg goose – force-fed so that its liver could be later used for pâté – was invoked as an image for the crammed examinee. Cramming and bird-stuffing were even given the playful term "ornithopachynsipaedeia" by James Wilson, that same senior wrangler of 1859 who broke down after the Tripos and forgot all the higher math he had learned.[26]

When the undigested food encountered an exam, a further metaphorical relationship was brought forth. The memorized facts were *regurgitated,* "vomited forth upon the examiner."[27] Huxley complained of boys being "stood up in rows and crammed like turkeys"[28] (birds again) and remarked on "how those dogs of examinees return to their vomit."[29] Huxley associated knowledge with food throughout his life. As a sixteen-year-old studying for exams in 1841, he told himself not to "cram a crude undigested mass in my head."[30] In 1877, now a fifty-two-year-old organizer of science teaching, Huxley spoke of candidates "as heavy and stupid from undigested learning as others are from over-fullness of meat and drink."[31] Indeed, Huxley's distaste for cramming motivated him to found his teaching laboratory at South Kensington. While such methods were supposed to show new science teachers how to better teach science, the laboratory experience was also intended to stop so many teachers from cramming their students from books: Huxley's dislike of learning science solely from books is well known.[32]

There were still other metaphors for exam preparation. Sometimes "cramming" was associated with contents jostling inside one's head – if one looked inside one would see no organized book of knowledge, but something resembling an old curiosity shop. There, in one corner, was a Greek verb struggling with a German noun; over here, in another, were thousands of partial fractions and intermediate equations riding roughshod over shelves full of bottles of chemical preparations.[33] This metaphor implies that the knowledge is not fully integrated, though not harmful. Another metaphor was "grinding," often to denote preparing for medical exams and sometimes those in Classics, and presumably implying dull but focused hard work.[34] Still another was construction. The Civil Service Commissioners noted above described "cramming" as raising a "showy superstructure without laying the necessary foundation."[35]

But the metaphor of foundations and superstructures could be used in more complimentary ways. William Briggs, whose highly successful University Correspondence College will be described below, said that one required basic "bricks" upon which to build, and whether one liked it or not, these bricks had to be memorized. He rejected outright that there was any "cramming." Briggs's employee, H.G. Wells, saw "cramming" as a "magic word" implying a "seedy rogue" and his "gaping, attenuous customers" – but he too thought some memorization necessary for any study of a technical subject like biology. A field such as anatomy, with all of the names of different body parts, could not be understood without memorizing those names. This meant that "cramming" as memorization was employed even at such elite locations as the Cambridge Morphological Laboratories.[36]

Free Traders in Education

But cramming usually had negative connotations, and its practitioners – private tutors not formally affiliated with a college – considered themselves "coaches," not "crammers." Private tutors held an ambiguous position in Victorian education, but they appeared wherever there were exams. For the need to pass an exam created good business opportunities. Smith's Establishment for Young Gentlemen, featured at the start of this chapter, was one such preparatory business. We have seen how at Cambridge, anyone wishing to gain a high place on the Tripos had to work with a tutor. Andrew Warwick has observed that in the early 1830s, tutors came to be known as "coaches"; they were likened to skilled drivers guiding a "team" of undergraduates quickly along a predetermined course of study. The best ones, like William Hopkins and Edward Routh, would maintain their reputations by carefully selecting the students they would tutor.[37] Such coaches taught tricks they knew would please certain examiners: one advised his students to use the "same notation" on the exam as one examiner did in one of his books. They were also to quote that examiner as an authority. It was claimed that out of such apparently minor changes, wranglers were made.[38] Over at Oxford there were also numerous coaches: Frederick Temple was considered unusual for *not* working with one, yet still doing well on his exams. Before he went on to administer the City and Guilds examinations, Phillip Magnus built up a business as a mathematics and physics coach for the College of Preceptors exams and for

London's Matriculation exam; his success meant he could set his own fees and choose which students he worked with.[39]

Greater challenges were faced by coaches working with less-brilliant students. At Oxford, where such "pass men" were said to "dominate the university," the "pass coach" was known for providing "a system of mechanical contrivances for retaining knowledge"[40] – enabling a candidate to provide a simulacrum of learning necessary to pass the required exams. At Cambridge, the hoi polloi, or poll-men, similarly outnumbered the honours students. Such students nonetheless also needed to pass their exams. The struggling student, according to Leslie Stephen, shunned the "condemnation of the examiner as the slave does the whip of his driver, and thinks all arts fair by which the attention of his enemy can be eluded." Many did not do well on tests. Stephen recalled one who failed his preliminary "Little Go" exam seven times, each time failing only one of the necessary papers – likening this student's mind to the small hand of a child trying to pick up six marbles when it could hold only five.[41] As at Oxford, similar pass-coaches existed at Cambridge. To assist their students, the coaches used different aids. They displayed huge sheets of paper shaped like a compass to help their charges memorize Christian principles; they used toy pulleys and inclined planes that demonstrated basic physical laws likely to be asked in the exam; they assigned catechetical books that "boiled down human knowledge into shreds and patches."[42] One mnemonic designed to answer fifty possible questions on Archbishop Paley's chapter on miracles ran:

> Posterior ages – distance climes;
> Transient rumours – naked rhymes;
> Particular – otiose assent,
> Affirmance of allowed event.
> False perception – some succeed,
> Some are doubtful – thousands feed.

This rhyme was produced by a fictitious character in the story "Mr. Golightly," a poll-coach known as "The Whopper" who made £2,000 a year drilling such mnemonics into anxious minds. Fictitious or not, the Whopper's mnemonic seems effective: the reader is invited to compare it with Paley's chapter on miracles.[43]

Although private coaching was often decried as "cramming" and its practitioners as "crammers," many rejected this label. Henry Moseley testified that it was unfair to disparage tutors as "crammers" because it was their steady focus on the single goal of preparing students to pass exams that made tutors effective.[44] Where some eminent medics called them "crammers,"[45] other medics insisted that tutors excelled at teaching out of necessity – for unlike professors, private tutors actually had to be understood by their students if they were to stay in business.[46]

The most successful private coach of the Victorian era – and one who rejected the label of "crammer" – was Walter Wren. Proprietor of Wren's Collegiate Establishment for the Preparation of Candidates for All Competitive Examinations, based in Powis Square, London, Wren specialized in tutoring candidates for the Indian Civil Service exams. His establishment opened in 1869, and over the next twenty-two years he prepared 427 (52 per cent) of all 827 successful candidates, with one of his students ranking first in twelve of those twenty-two years.[47] Wren's students attributed their success not to mere tricks but to some of the very moral values that examinations were supposed to indirectly assess: self-discipline, hard work, and attention paid to specific topics. One student later recalled that he never knew what "real work" or "good teaching" involved – either at public school, or later at Oxford – "till he went to Wren's." This claim was supported by the account of another private tutor, who estimated that his own students studied nine hours a day, compared with five for the average Oxford student. Another of Wren's students suggested that the coach's secret was to ensure his students learned a few subjects very well rather than scatter their attention widely. "Little and good" defined his system: idle students were asked to leave. Wren frequently examined students not only to assess their knowledge but to familiarize them with being tested.[48] As Wren's reputation spread, candidates travelled from India to London to study with him. They would write the civil service exam and usually succeed on it, and then return to India as civil servants. Other candidates studied with Wren too: Bertrand Russell studied at Wren's before going "up" to Trinity College, Cambridge.

Wren was a Liberal MP for a short time,[49] and like other Liberals he justified his teaching methods by invoking political economy. Public schools were "inefficient," he said: they were careless, emphasized sports far too much, and had excessive holidays. By contrast, the only goal of Wren's Establishment was exam success.[50] By ensuring his students knew the material for an exam, there was no gap between knowledge and its symbol; other considerations could be deemed irrelevant and discarded. (When one mother complained that that her son did not go to church while attending Wren's, Wren replied that his business was to "fit your boy for the Indian Civil Service, not for the Kingdom of Heaven."[51]) He charged that the public schools were actively harming the cause of education by lobbying for standards to be changed to fit their system, rather than the other way around. Likening their efforts to trade unions seeking monopoly control over a market, Wren fought back.

The successful coaching of Wren and others turned the Indian Civil Service (ICS) exams into a field resembling the race of the Red Queen. The percentage of privately coached candidates more than doubled between 1865 (33 per cent) and 1874 (84 per cent).[52] Questioned about this rapid expansion, Mark Pattison, rector of Oxford's Lincoln College and an ICS examiner since 1863, dolefully noted that the ICS examiners had sought to "defeat the system" being set up by coaches – but they hadn't been very successful.[53] There were too many coaches. In 1862, one pseudonymous letter-writer to *The Times* complained that someone in the Civil Service Commissioner office must have been forwarding ICS applicants' names and addresses to exam coaches, since after registering for the ICS exams he had received thirty or forty letters, often lithographed, from different coaches all promising to teach him how to succeed on the exams.[54] Such Victorian "junk mail" shows that free trade in education had indeed arrived.

Learning and Earning

For their part, national school teachers were encouraged by one guidebook to pay close attention to how inspectors examined students during their visits: their "slightest act" was "significant." Despite attempts to ensure uniformity, inspectors did show differences: in geography, for instance, some usually asked about oceans while others asked about land masses. Hence the guidebook recommended learning the identity of the visiting HMI in advance and studying his preferences. Moreover, although taking notes during an HMI's exam was forbidden, teachers ought to recreate the exam from memory afterwards and pay attention to other inspector patterns, such as the pace of reading or the speed at which arithmetic questions were asked.[55] Some teachers went further: for instance, in 1879 those in the Liverpool area formed the "Checkmate Society" – a "defensive 'league,' to prevent, by all means short of fraud, failures and losses at examinations." In addition to comparing notes and compiling from memory each HMI's examinations, teachers would forward their notes to other Society members whose students had not yet been visited and examined.[56] By 1884 such routines had become even more open – the editor of the *Practical Teacher* was actively soliciting and publishing questions from national school teachers whose students had been examined.[57]

Further complicating exams were various attempts to exploit loopholes in the rules. The more money at stake, the more such gaming occurred. Once schools and candidates discovered loopholes in the rules and exploited them, the governing institution reacted by changing the rules. Hence there occurred arms races of reactivity, especially when rules were vague. For instance, the original 1862 Revised Code allowed each national school student to be examined once on a single standard. But it gave only ranges of ages. Soon there arose a practice in which a student, after passing a higher standard in one year (e.g., an eleven-year-old boy in Standard 4) was presented for testing in a *lower* standard the next year (e.g., the now twelve-year-old in Standard 3). In response, in 1870 the Education Department changed the rules.[58] Over time, stratagems grew more complex. One 1885 guidebook suggested that since some students would always do poorly on exams, each teacher should place them on a "blacklist" to avoid their being examined: the student might claim a health exemption, for instance. Failing this, it might be necessary to create a reason to expel them. But teachers should beware of other teachers' blacklists and watch out for "an incursion of strange children" into their own classes just before exam time.[59]

The complicated rules of the DSA were also fertile grounds for gaming.[60] At first there was no limit as to how many exams a candidate could write, so numerous candidates were sent to take a "shot." Nor was there an age limit, so some students as young as nine were sent to take the exams.[61] In response the DSA plugged the loopholes by changing rules or even retroactively disqualifying certain candidates. To enforce its rules, the DSA hired Royal Engineers after 1867, its own cadre of inspectors.

Someone who had grown up exploiting loopholes in DSA examinations, and who then taught others how to exploit them, was H.G. Wells. In the years before he wrote *The Time Machine* in 1895, he made money first by taking, then coaching for, the exams of the DSA. His accounts reveal how the examination cash nexus that emerged in the mid-1850s had evolved by the mid-1880s. Wells wrote his first exam at age twelve, when he tied for first place on a nationwide bookkeeping exam for the College of Preceptors. After a short time as a draper's apprentice, and acting as a pupil-teacher for an uncle, the need to learn Latin for pharmacy school meant he was sent to Midhurst Grammar School in February 1881. Its headmaster, Horace Byatt, noticed Wells's ability to learn rapidly "just by reading" and decided that Wells would be an excellent candidate for DSA exams.[62] By May of that year Wells wrote ones in animal physiology and physiography and gained first-class results on both. After a second stint as a draper's apprentice, he returned in August to act as Byatt's assistant. At age seventeen Wells was paid £20 a year to teach and learn at the school – "All I have to do now is pass examinations not to flatter employers shopwalkers & heads of departments," he told his mother.[63] Like his doppelganger in his later *Love and Mr. Lewisham*, Wells pushed himself, awaking at five in the morning to study; after teaching at Midhurst, he went to evening DSA classes, collecting various "bluish-green South Kensington [DSA] certificates" in subjects from astronomy to physiology.[64] By the spring of 1884 Wells was studying for a second round of DSA exams, presumably the advanced papers. He earned five honours results, even winning a Queen's Prize in Human Physiology. A delighted Byatt offered to double Wells's salary as a pupil-teacher to £40.[65]

Wells, however, had other plans. Anyone receiving honours on a DSA exam was informed of the teaching school scholarships, and by September 1884 Wells was in London, paid to study in the red-brick and terracotta training college building, taking biology and zoology in a course taught by his idol, T.H. Huxley. Unfortunately, Huxley only ever taught one class, becoming ill in the spring of 1885, and Huxley's demonstrator, G.B. Howes, took over.[66] But Wells's first year at the college went well, earning him honours in elementary biology, mathematics, and advanced zoology. In January 1886 he passed the University of London's Matriculation exam, intending to obtain its bachelor of science. But he became distracted by women and socialism, and failed two courses: his scholarship was not renewed in June 1887.[67]

So Wells returned to teaching. It was a world of intense competition: too many certificated candidates chasing too few jobs. A job at a north Wales school he left after five months, and Wells supported himself by tutoring candidates for army exams. By 1889 he had been hired at £60 a year to teach science at J.V. Milne's Henley House School: more precisely, Wells's job was not so much to teach science as to teach his students how to pass the DSA exams in science. Despite his experience in Huxley's teaching laboratory, Wells did not have his students conduct classroom experiments; he reasoned that experiments wouldn't help his students pass exams. Instead, they copied out his coloured-chalk blackboard drawings to improve their "illustrating examination answers." His *New Machiavelli* lampoons how DSA classes could be run at the "very highest level of grant-earning efficiency," no doubt drawn from Wells's own experiences. Lessons included not only the dictation and copying out of model exam replies, but the students' memorizing "exceptionally difficult but grant-earning words, such as 'empyreumatic' or 'botryoidal.'"[68]

Wells continued to write exams himself. In July 1889 his "state of cram" got him a second-class result in the zoology paper of the London BSc intermediate science exam. He calculated which subjects would lead to honours results but take less effort: better to write a thesis on mental and moral science that was "decidedly inexpensive" in comparison to lab work on crystal structures. By December he charged classmates one guinea for fifteen hours of coaching. He also studied for the licentiate exam of the College of Preceptors – a bet that paid off when he became the only one of 108 candidates to gain its full diploma, as well as gaining £18 for his papers in natural science, mathematics, and the theory and practice of education.[69] These victories Wells took to Milne when he asked for a raise and fewer work hours – otherwise he would go to another school.[70] Such bargaining worked, and with his newly freed-up hours Wells gained a side job as zoology tutor for the University Correspondence College (UCC), spending twenty hours each week (making £2) marking students' mailed-in answers. The UCC's proprietor, William Briggs, offered Wells a permanent and better-paying job there when he received his bachelor of science. The higher the grades he received on his exams, the higher his salary would be, since "certified triumphs" were important for the UCC's image.[71] With mercenary humour Wells celebrated Briggs's offer to a friend – "STOCK EXCHANGE INTELLIGENCE GREAT BOOM IN WELLS," a "surprising run" on the Wells's concern.[72] By November 1891 he had left Milne's school to work full-time for the UCC, working fifty hours a week at two shillings and six pence an hour: enough to finally get married.[73]

The University Correspondence College emerged in the aftermath of the University of London's Clause 36 enabling distance education. It was Clause 36's shadow, its dialectical partner, its opponent in the ensuing arms race. Correspondence teaching had begun in the 1850s, with Pitman's Shorthand and Pitman's Metropolitan School, whose motto was "Learning and Earning."[74] After informally coaching friends by mail for their University of London mathematics and chemistry exams in the early 1880s, Briggs officially founded the UCC in 1884.[75] By 1887 every one of its students had passed London's Matriculation, and UCC advertising declared that no student had ever failed a London course if they had previously taken the relevant UCC preparatory course.[76]

UCC's triumphs were due to Briggs's reverse-engineering of the University of London exams.[77] To get UCC students to pass, Briggs – trained as a chemist – deployed the logic of chemical analysis and synthesis. Just as there were a finite number of elements that could be combined into more complex chemicals, he reasoned that there was also a finite number of possible topics in any subject: a periodic table of examinable knowledge. All exam questions were merely "combinations and permutations" of these simpler topics and questions, regardless of whether the course was in Classics, mixed mathematics, or law. Briggs's logic was thus one of partition. His method was simple: because London's exams were intended to create a predictable standard of attainment, and because they were always published, Briggs collected all of London's past syllabi and exam questions. Then he broke up each question into its constituent elements, tabulated them, and compiled all of the material he deemed necessary to answer each of these simplest questions. He divided the material into thirty equal sections: each one became a "class" or chapter of a UCC textbook. Each chapter ended with model questions and answers.[78] The London exam simply became the thirty-first set of questions to answer.

Wells was the fortieth tutor Briggs hired for the UCC. Although Wells had not yet used the practical lessons learned at Huxley's teaching laboratories, he was hired because of this lab experience in zoology. Lately, the University of London had been emphasizing the value of practical examinations, and the UCC needed to adapt by hiring a zoology tutor. At first, Wells taught practical bench-work such as dissection in a little room in Briggs's University Tutorial College, housed in quarters above the UCC's bookstore off the Strand.[79] By 1891, however, the Tutorial College had moved to a brand new 10,000-square-foot building in Red Lion Square, Holborn; its biology teaching laboratory resembled "a billiard saloon," containing four enormous teak-topped benches, ventilation, telescoping lights, and diagrams with coloured chalk.[80] Wells taught six to thirty-two students in classes totalling forty hours, either twice a weeknight over several months, or in a seven-day period during the holidays. He began each class with a demonstration dissection, then wandered around the laboratory to watch others' dissections. Apparently he was an excellent teacher, coupling directness with a slight cynicism, and he showed extra sympathy to impoverished students.[81] Although the college had set aside two rooms for women, Wells's lab was integrated, and it was in one of his Tutorial College classes where he first met Amy Catherine Robbins, for whom he would leave his first wife, Isabel. Robbins would redraw the diagrams for the second edition of his first major publication, the *Text-Book of Biology*.[82]

Wells's *Text-Book* used Briggs's insights in crafting what Wells would later call his own "examiner-defeating mechanism." The book's goal was to help the reader pass the London intermediate science exams in zoology and botany.[83] The *Text-Book* combined Briggs's partitioning logic with Huxley's use of specific "type-specimens" – for instance, frog, rabbit, dogfish, crayfish, earthworm, and the freshwater *Hydra* polyp – each of which instantiated some basic anatomical or physiological principle. Each of its thirty chapters ended with three or four questions, closely resembling previous questions on London's zoology exams.[84] A new student would sign up with the UCC and pay three pounds, three shillings for the course in advance, read a chapter of the *Text-Book* a week, and answer its questions, which would then be mailed to UCC headquarters in Cambridge. The headquarters forwarded the responses to Wells, who in red ink marked each answer and made comments such as "You must read number 35 again" or "The matter you have introduced here is not required for a pass."[85] Wells mailed them back to the UCC, which returned the individual workbooks to the students. Thus practised, when students took the real London intermediate zoology exam, they saw it as their thirty-first paper.[86] For students unable to attend Wells's laboratory to gain practical experience, the *Text-Book* showed them how to dissect and suggested where to obtain specimens: butcher's shops and ponds, for instance.[87]

The UCC made Briggs wealthy. The University of London's own examiners certainly had no love for it. In July 1893 when a UCC student at a chemistry exam was overheard praising the college, the examiner sarcastically responded, "Oh, if *you're* satisfied, that's all right." The student complained to London's registrar about this exchange.[88] Yet Briggs's methods were exceptionally successful. For his part Wells recalled how successful his UCC tutoring was: in one year all of his candidates took first class on it and thereby drove down into second class all the students of his rival coach – Edward Aveling, son-in-law of Karl Marx, who was a more experienced coach and textbook author than Wells.[89] Wells even started to encounter a new kind of pupil – medical students bored by their hospital lectures. But most of his students were still prospective teachers or headmasters seeking "to add B.Sc. to their caudal adornments" and thereby advance their careers, just as Wells had done.[90]

Wells later saw his work as a tutor being fraught with "comedy and absurdity": libel considerations prevented him from writing a comic novel about Briggs, *Mr. Miggs and the Mind of the World*. Publicly he complained that his coaching was marked by a *failure* to educate, providing only examination results that looked like an education. "In the true spirit of an age of individualistic competition, we were selling wooden nutmegs or umbrellas that wouldn't open, or brass sovereigns or a patent food without any nourishment in it,"[91] he lamented. The images of counterfeit currency and false food were used here, again, to convey the danger of too much focus on attaining credentials. Yet despite this public criticism, we know that Wells still privately believed in the benefits of a widespread examination system: in 1921, to the visitor Chen Yuan, Wells spoke favourably of the Chinese imperial examination system and suggested its adaptation to new topics.[92]

But of course Wells's coaching actions followed the letter, if not always the spirit, of the rules intended to keep examinations on a level grade. Others – far more single-minded and unscrupulous than Wells and Briggs – were not so constrained.

8 Immoral Economies: How to Cheat on a Victorian Exam

And the examinee
Took a paper in hand;
It was Physics, the same
He did not understand;
Yet he smiled on that faint purple writing
With a smile that was childlike and bland.

For on it he found
(To relate it I grieve)
Sundry questions on sound,
And he had in his sleeve
Mr. Stone's most compact little treatise,
The same with intent to deceive.

A beautiful sight
Was the smile on his face,
As he set off to write
At a terrible pace:-
But it changed when a glimpse of the treatise
Had brought Mr. K ---- to the place.
...

In the scene that ensued
I did not take a hand;
But the table was strewed
With (produced by command)
The tips he'd prepared for the paper,
In case he should not understand.

From the case of his watch
Notes on optics he brings;
Lists of colours that match
And some similar things;
Whilst neatly inscribed on the dial
Were the laws of vibration of strings.

His wristbands display
What you'd little expect
That's the spectrum of K.
(Though it proved incorrect)
As well as the series of Seebeck
And remarks on the Peltier effect…

– Ova [pseud.], "The Examinee: A Reminiscence."

On Friday, 17 May 1878 at 3:10 p.m., the science teacher Robert Edward Hemblington Goffin was charged with defrauding the Department of Science and Art on a massive scale. Earlier that morning his accuser, the DSA's science director John Donnelly, had received evidence that Goffin had gained early access to its exams and drilled the answers into his teenage students at London's United Westminster Schools before they wrote. Donnelly had suspected Goffin of fraud since 1865; he had even confronted him in 1872 over it. Now he held the evidence in his hands: student notes of a review session held by Goffin, penned six hours *before* the DSA exam in inorganic chemistry was to be written, containing answers to questions on the exam. Donnelly rushed to the United Westminster Schools with two other DSA inspectors. First they asked to speak with Goffin in an empty classroom, and there, accused him privately. Confess, Donnelly said; reveal the source of the leaked questions, and he'd be lenient. But Goffin denied all. "What questions?"

Now the situation became deeply unpleasant. Huxley once described his friend Donnelly as hot-tempered but quick to recover; that afternoon Donnelly was wearing a white bandage "around his face" – possibly for a toothache – and seems to have been especially irritable. Goffin's students were hauled into the empty classroom for individual questioning. Some confessed, with details in their stories corroborating one another. Donnelly then decided to bring forty or fifty students all at once into the room for questioning, but unwisely allowed Goffin to remain in the room. Unsurprisingly, some of the confessants now recanted. No, Goffin had not given them unfair assistance. What questions? By about 4 p.m. what had begun as a discreet visit had become a fiasco, with Donnelly stamping his feet and calling the students to their faces "wretched, miserable boys who are systematically taught to lie."[1] It was not the proudest moment in the career of this decorated veteran of the Crimean War, now a lieutenant colonel.

Donnelly's disastrous visit sparked what became known as the "Goffin Affair." It resulted in a parliamentary inquiry, a precedent-setting court case, and even a House of Commons debate between Goffin's critics and allies. Such a case is important to understanding examination misconduct if only for evidentiary reasons, since successful misconduct leaves no trace. Since Goffin was caught, the resulting documentation from this affair – particularly testimony sometimes drawn out unwillingly, from witnesses under oath – becomes a valuable historical resource that sheds light on examination misconduct and other forms of Victorian academic fraud. We glimpse the shadowy underside of a new world made possible by widespread examination – a world populated by teachers and exam candidates who felt unbound by the moral economy of examinations. What Mary Poovey calls the "problematic of representation" describes the gap between a sign and its referent, such as currency and its real value, and she is interested in what happens when the gap grows too wide.[2] Such a notion is applicable to examinations and examination misconduct, too: a gap between on the one hand signs of knowledge, competence, or skill (such as a person's credential), and on the other hand the actual grounds for that sign.

Yet it is important not to see the issue as merely involving individual lapses of "academic integrity." For examination cheating can also be systemic. Although the standardization of examinations had made it possible to assess on a much larger scale, so too had this standardization made possible fraudulent misrepresentation on a much larger scale. Institutions also have incentives to cheat: a form of reactivity. Metrics can be more easily manipulated in an institution's favour because such numbers, as abstract representations of a situation, are easier to change than that situation itself.[3] It takes less effort to underreport the crime rate than it is to actually cut crime. Such incidents have become so common – in fields as varied as crime, finance, economics, or the environment – that they are explained by "Campbell's Law," named for the social scientist Donald Campbell. Campbell's Law holds that the more a single set of "metrics" is relied upon to make decisions, the more likely it is that those numbers will be altered corruptly, distorting the very processes they are supposed to neutrally monitor.[4] The Goffin Affair can be seen as an early, albeit extreme, case of "juking the stats."[5]

Certification and Collusion

In 1859, just as Donnelly and Cole were putting the finishing touches on the DSA's new system, Goffin visited the Reverend Horace Noel in Exton, Oakham, near Grantham, bearing glowing testimonials about his teaching abilities. As chair of the board overseeing a national school, Noel hired Goffin as a schoolmaster.[6]

The *Popular Science Journal* trumpeted the DSA's system of payment by results as raising both the "intellectual condition of the masses" and the income of teachers, especially female ones.[7] Goffin seems to have realized this earning potential almost as soon as the DSA system was established, travelling to London in November 1860 and passing Tyndall's DSA teachers' exam in acoustics, light, and heat, gaining a certificate in that subject. This certificate not only entitled Goffin to money if any of his students passed their exams in acoustics, light, and heat, but also £10 a year, presumably to buy teaching materials. These payments would be administered by the same organizing committee that also administered the DSA examinations.[8] Goffin soon gained certificates in at least eight other subjects, including pure mathematics, theoretical mechanics, magnetism and electricity, and animal physiology. He was a gifted exam taker, earning the gold medal in inorganic chemistry one year.[9]

Goffin's new DSA certificates became important for his income when in 1862 Lowe's Revised Code drastically reduced the amount of money he could earn as a national school teacher. Moreover, these certificates made him more attractive to other, larger schools. The contrast between the opportunities offered by the Revised Code and the DSA examinations is shown by two 1862 job openings at the Liverpool School of Science: one, the position of Assistant Secretary, paid only £20 a year for two nights' work a week. It attracted 400 applicants, mostly teachers wishing to supplement their salaries. The other position, a full-time lectureship in geology, was given to the candidate with eleven DSA certificates and a demonstrated ability at producing medallists.[10]

It was in May 1865 when Goffin first came to the attention of roving DSA inspectors. His students had all done well on their various exams, and so the inspectors dropped in to ask viva voce questions of the students, individually. They found that the students all made exactly the same mistakes as they had on their exams. In his defence, Goffin pointed out that he used textbooks that reprinted past exam questions, and that DSA exams tended to fall into predictable patterns: this must have been the cause of such coincidences.[11] Nonetheless, in October he resigned from the Exton school and stopped teaching DSA classes.

Accounts differ as to why: the DSA later claimed he left because of their suspicion about exam misconduct, whereas Goffin blamed a conflict with Rev. Noel's son over a failed investment. He and his family moved to Essex, where Goffin taught at the Purleigh parochial schools. When he declared bankruptcy in 1866,[12] such financial troubles focused Goffin's mind on teaching only the more valuable DSA science classes.

In June 1868 Goffin moved to St John's School in Woking, Surrey, as science master, and set up DSA classes there. By May of the following year he had fifty-three science students writing exams. He was helped by his sister Jane, who had herself won a DSA medal in mineralogy.[13] Goffin expanded his teaching the next year, taking on an additional evening class of forty-seven at the Guildford Town Hall and Mechanics' Institute; he was even recommended to Guildford's mayor by a DSA inspector, Royal Engineer Captain Edward James, who either overlooked or was unaware of his own department's concerns about Goffin. For James, all of the students' exam successes were sufficient evidence of Goffin's teaching skill. And in that year, 1869, Goffin's Guildford students won ten prizes, and thirty-three pounds and fifteen shillings; his Woking students earned two prizes and thirty-six pounds. For her own teaching, Jane's students earned twenty-two pounds and fifteen shillings. Some portion of the DSA money would have gone to the school and the Mechanics' Institute, depending on what Goffin had negotiated with administrators, but both he and Jane likely would have kept a substantial portion of this money.[14]

Shortly thereafter Goffin took on a new assistant at the Guildford Mechanics' Institute: Charles Ledger, whom we first met in Chapter 3 answering Huxley's questions about the trachea. Ledger received additional instruction from Goffin in the evenings, and passed seven DSA exams in May 1871. In May 1872, Ledger wrote nine. Another of Goffin's students wrote even more – twelve exams. One nine-year-old student wrote six, marking a busy year for the spread of scientific knowledge in Surrey.[15] One reason why Ledger, along with Goffin's other students, had performed so well was that Goffin knew what the questions would be in advance: that year, Ledger claimed to have seen questions from the upcoming mechanics exam sitting on Goffin's desk. Goffin then forwarded the exam questions to a teacher at a nearby school, wishing to also claim these teachers (and their DSA payments) as his own students when they wrote the magnetism exams. Although students were supposed to attend at least twenty-four one-hour classes, Goffin got around this by falsifying the attendance registers.[16]

Goffin's May 1872 successes drew more DSA attention. In early June inspectors questioned him and his St John's students. When on 20 June,

Goffin came to study in T.H. Huxley's new South Kensington teaching laboratories, Donnelly learned of his arrival. He strode into the laboratory and "in the presence of a number of students, abused [him] in the most violent manner upon the question of cramming." Although Goffin replied that there was no rule against students taking so many exams, the DSA still cancelled all of its 1872 payments to him, and then changed its rules to lower the number of payments that could be made on the exams written by a single student per year: five.[17] Ledger was flagged as one of Goffin's students, which may be why he failed the next year's physiology exam, as described in Chapter 3.

Cribbing and Copying

As well as the financial benefits, Goffin probably justified his behaviour on the grounds that DSA exams were empty rituals that did not actually reward merit and knowledge. Such a rationalization seems necessary for exam misconduct. The point is borne out by the fact that misconduct almost always occurred on pass exams – the Cambridge ones assessing knowledge of Euclid or evidence of God's handiwork in nature were often seen as empty, shown by the cynical Cambridge proctor who turned his back on the hoi polloi examinees. By contrast, competitive exams saw very little misconduct. On an abstract level, this is likely because most examinees shared the moral economy of an exam and its premises. More concretely, there was less misconduct because a candidate could gain a higher place only by forcing competitors into lower places. When one candidate saw another candidate cheating on a competitive exam, it was always reported.[18]

Of course, copying off of others or bringing in hidden aids often backfires on the test-taker because their illicit information is wrong. The 1883 poem by "Ova" that began this chapter satirizes a candidate's intention to cheat by ingeniously hiding information – but most of it is incorrect. In the fictitious *Adventures of Mr. Verdant Green* the eponymous character meets a fellow Oxford student, Mr Bouncer, who has created an elaborate coat he wears to exams. It has hidden pockets and built-in notes, some of which pop out with pulled strings, hooks, and wires: three sets of cards pop out with answers to Euclidean questions (in the same order as the *Elements*, no less). Yet Bouncer fails not because his aids are discovered, but because the propositions he has copied from Euclid are incorrect.[19]

Both accounts are fictitious, of course: writing about cheating is hindered by the evidentiary problem that only unsuccessful cases ever get discovered. Nonetheless, one can find some accounts of copying and "cribbing," or smuggling in notes. In April 1852 Henry Labouchere, son of Lord Taunton and student at Trinity College Cambridge, was accused of smuggling in notes for his previous examination at the Senate House. Labouchere claimed that he was writing a note to his neighbouring friend asking how he had just answered one of the questions in scriptural history, but this weak explanation was found unconvincing and Labouchere was suspended for two years.[20] In 1855 *The Times* drily noted that at one January entrance exam for the military academies, an "active and obliging candidate" had worked out the mathematical papers for all the surrounding candidates – making their success less the result of industry and competence, and more to their "accidental propinquity to this central source of light and information."[21] One 1862 College of Preceptors exam saw copying at one country school, but the culprits confessed and received no certificates. At the 1866 run of the Cambridge Local exams two junior students were caught copying from one another, and upon comparing answers afterwards the examiners discovered suspiciously similar answers among another eleven candidates. In response, organizers devised new procedures: they held exams in larger rooms, or marked the exams in the order in which candidates sat, facilitating the detection of identical answers.[22]

Because communication over long distances leaves paper records, cases of misconduct on overseas exams are easier for the historian to find. The June 1880 Matriculation exams of the University of London, arriving from Mauritius, underwent a near-comical level of scrutiny because of a suspicion of copying. All nine candidates gave near-identical answers to the chemistry and mathematics portions of the exam. The ensuing report found that the same solutions, same diagram letters, and same order of proofs were used, and its authors went on to speculate about who copied from whom and whether some candidates had copied from a single piece of paper. Although the senate resolved to hold no further exams in Mauritius until the Royal College showed it had ended the copying problem, the senate must have been easily satisfied quickly, as London's BA exams were written there that October.[23]

Although practical exams were supposed to be a check on merely "crammed" knowledge, misconduct on those was also easy. The Royal College's 1873 request to hold University of London BSc exams in Mauritius was rejected because there was no way for London's examiners to supervise the practical exams there.[24] Twenty-five years before this Mauritius request, William Thomas Brande, who was running a chemistry exam at the university, complained that because tables at his recent practical exam were set too closely together, candidates intentionally gave hints to other, "less shrewd" candidates under the pretext of noting results, or by loudly asking extremely specific questions of the supervising examiner.[25] Complaints like Brande's led to a campaign for larger examination rooms for the University of London: until the 1860s the Matriculation exams were written in the east wing of Burlington House, which was too small for the number of candidates writing therein. In April 1856 London's senate complained to the treasury that "irregularities on the part of the Candidates cannot be prevented" unless the examination room was large enough and laid out so that every candidate could be watched.[26] Eventually, twenty square feet became the prescribed amount of space for each examinee at Burlington House. Conversely, however, too much space could lead to cribbing, as the Rector of Exeter warned in 1872 when new schools at Oxford allowed thirty square feet per candidate.[27]

Returning to the DSA exams, security concerns grew alongside their popularity. In May 1867, the piecemeal use of Royal Engineers as roving DSA inspectors was formalized. And so it was in this year that problems started to be discovered, such as copying. The DSA's carefully specified exam procedures had been ignored in Belfast, Carrickfergus, and Galway. The violations were blatant: at one of these three locations the exam package was opened early and a blank copy passed to a teacher waiting outside the room.[28] The teacher rapidly answered all the questions and passed the script back to an examinee, who then consulted the copy and passed it along to others. The DSA cancelled the exam results from all three centres and revoked the right of the teachers involved to earn its money for three years. It also developed countermeasures: not only a more secure envelope to hold the exams, but also the amalgamation of numerous small testing locations into larger ones, facilitating inspection. These measures seem to have worked: by 1875, with a 40 per cent cut in the number of Irish testing locations, reports of copying there declined.[29]

"Reduced to a System"

Although Goffin had his 1872 DSA payments withheld, we have seen how the Department of Science and Art treated him inconsistently, with some officers publicly insulting him and other officers providing reference letters that praised his "efficient" science teaching and his success at training gold medallists. Goffin furnished these testimonials when in January 1874 he applied for, and obtained, the headmaster's position at the new United Westminster School (UWS) in London.[30] Built on the site of St Margaret's Hospital (it is now Westminster City School), its science teaching laboratories were especially well equipped.[31] Goffin, having declared bankruptcy only eight years before, now earned £500 a year. A further indicator of Goffin's negotiating strength is shown by a clause in his contract: the board of governors of the school could not hire or fire new teachers without consulting him first. Meanwhile, Goffin's opinions about effective science teaching were being respectfully listened to at panels (he opined that many who spoke about science education actually "knew very little ... of the school-room").[32] One problem loomed, however: in the fall of 1873, around September, the unwed Jane Goffin had become pregnant. Goffin's apprentice, Charles Ledger, was the father-to-be.[33]

The first classes of the UWS began on 20 April 1874, with one hundred boys. Eleven days later, Jane's child was born. Ledger went to live at Goffin's mother's house, and the couple married in March 1875. Goffin invited Ledger to apply for a teaching position at UWS, telling him precisely what to write on his application letter to the board of governors. He noted that they would train students for examinations in the same way as they had at the Guildford Mechanics' Institute and St John's School. Ledger was appointed in April, at £100 a year, his contract specifying £5 raises each year to a maximum of £130. The school added about a hundred students a year until by 1880 there were 667. New buildings were constructed to accommodate the rapid growth, which the board attributed to Goffin's organizational skills.[34] Also growing were Goffin's DSA payments. In 1875, the first year in which DSA exams were held at the UWS, Goffin made an additional £86; in 1876, an additional £355. He received the money from the DSA indirectly – the money first went to the School Science Committee, headed by the chair of the board, Liberal MP Sydney Waterlow; the committee then paid the money to him.[35]

Later witnesses unaware of Goffin's secret believed that he had a preternatural ability to predict which questions would appear on the DSA exam. The truth was more prosaic. When in 1877 the UWS pupil-teacher John Coates took the elementary DSA exam in theoretical mechanics, although he had not taken the required twenty-four classes (the register was falsified again), he was prepared by Goffin privately in advance just as Ledger had been. Goffin gave Coates a succinct definition of how levers worked, using diagrams – and later that evening the exam duly featured a question about levers. When the DSA inspectors came around, Coates was kept away from the school in case he was in a confessional mood. Goffin made £386 on DSA results that year, added to his salary.[36]

At home, however, problems were brewing. For some undetermined reason Goffin's relationship with his sister and new brother-in-law deteriorated, leading to a poisonous dispute with the two on one side and Goffin on the other. Ledger escalated by taking their dispute to the board of governors, but was told by Sir Sidney that the board had "implicit confidence" in Goffin. Ledger did not speak to Goffin again, and in August 1877 he left the UWS to become a constable with the Metropolitan Police, resulting in a pay cut of about £20–30 a year.[37] He would furnish much of the evidence against Goffin while trying to foment unrest among his former colleagues.

No later than April 1878, given what we know of the DSA examiners' prescribed procedures, its examiner in inorganic chemistry, Edward Frankland, prepared the questions for that subject's three papers. He forwarded printers' proofs to and from the department in special security envelopes. Indeed, due to printers' leaks in other cases, the department had developed additional steps in 1873 to ensure security. Now, all exams were printed on DSA premises by a detachment of Royal Engineer printers; the exams were then placed in a locked room under guard. There were also other security measures: a mandated annual exchange of keys (implying a changing of locks each year), and passes required to remove parcels.[38] While such written orders are obviously not always followed, they do show the attention paid to security. Other places where exams were printed also had security procedures: at Oxford University Press the number of proofreaders was kept to a minimum to reduce the danger of leaks, and it put press-marks on the bottom of exam pages to hinder them being handed out at the wrong time.[39]

After printing, the DSA exams were mailed out, in the evening, two days before the exam. This additional time was included in case of postal delays. The exams were supposed to arrive the following morning and be received by a member of the United Westminster Schools organizing committee – but in this case the package arrived around 8 p.m. that same evening. The postal worker later recounted that the package was too large to fit in the school's mailbox, and so he gave the package to the school porter.[40]

At this point the evidence from both the parliamentary inquiry and the parallel internal inquiry of the UWS becomes tenuous, but it is probable that the school porter dropped the package of chemistry exams off at Goffin's on-campus house.[41] One must conjecture what occurred next to the exams held in their security envelopes, designed by a Post Office expert who had assured the DSA that they could not be opened without detection. A detailed imaginary account in an *Educational Times* editorial shows that someone had thought deeply about how one might get past the security envelopes:

> Closed doors, the application of a red-hot knife to the sealing-wax, the spout of the kettle to the gum, and the cover of the papers falls off; one is

> copied and replaced; the gum and wax are restored; all is as before, except that, for the final lesson before the examination, the students may have revealed to them the whole of the points on which the questions turn.[42]

Contrary to the Post Office experts' assurances that it could not be done, one of the DSA inspectors did learn how to open the exam envelope undetected, probably by steaming it; it took his colleagues ten minutes to figure out how he had managed it.[43]

On the afternoon of Wednesday 15 May, and again on Thursday afternoon, Goffin held review sessions for all students taking the elementary inorganic chemistry paper. At these sessions, his students wrote out notes on the atomicity of methane, sulphuric acid, phosphorus pentafluoride, and chromic fluoride (two students putting them in exactly that order), and drew graphic formulas of hydrochloric acid, sulphuretted hydrogen, and ammonia. He exhorted the boys to further study by noting the prizes and money that could be won. The notes of one attendee, Robert Jones, eventually made their way to a young UWS pupil-teacher intent on stopping Goffin's "system."[44]

At 7 p.m. that Thursday the exam was written; on the elementary paper were such questions as

> 10. Draw the graphic formulae of the following compounds: - Ammonia, water, sulphuretted hydrogen. Give the active atomicity of each element in the following compounds: - CH4, SO2HO2 [*sic*] [sulphuric acid], PF5, CRF6, and SO2.[45]

The pupil-teacher now had evidence that Goffin had taught in a half hour the material directly relevant to eight of the most valuable questions on the exam. Around 11 a.m. the following morning, he and his brother showed up at Donnelly's South Kensington office with the notes. We have now reached the scene where this chapter began – the botched investigation by Donnelly, Iselin, and Abney when they arrived at the school. At first the investigation was handled well – William Abney, elected Fellow of the Royal Society in 1876 for his discoveries in chemistry, individually questioned five students and established that at the pre-exam review session Goffin had discussed specific substances – for instance, methane, sulphuric acid, and ammonia – that would appear on the exam only six hours later. In his defence Goffin later argued that questions about the percentage compositions of certain substances always appeared on the exams – but this did not explain why the same substances in the notes were also asked about on the exam, sometimes in the same order. Frankland, who had drawn up the exam, claimed that the odds of picking the same substances, in the same order, were about the same as picking the winner of the Derby, out of thirty starters, for the next ten years. The boys' individual accounts corroborated one another, and Goffin's reaction became "exceedingly painful" to watch, recounted Donnelly.[46] But then his poor decision to allow Goffin to witness the questioning of the boys – leading to Donnelly's outburst – gave Goffin a crucial way to defend himself to his board of governors. He could claim the DSA had intended from the start to establish his guilt.

On 28 and 29 May several letters arrived at the UWS for Goffin. One was from the DSA announcing that the department had suspended his certificate and that no payments would be made to UWS students that year. Another read,

> Goffin, You are a scoundrel! – a seducer of your *pupils*, whom you ought to *protect* – a destroyer of *innocenci virtui;* but *vengeance* is on its way! You have had a foretaste, but all shall be brought to light, and you shall drink the cup to its dregs.[47]

The letter was in Ledger's handwriting; indeed, Goffin had somehow intercepted another letter from Ledger to Coates (whom Goffin had taught about levers):

> I pity you poor b------s in such a hell of a hole. I would rather be a good-looking prostitute than in such a place. I suppose the eyes look large and black, and as wild as the devil. Can't say any more; only look to see if this has been opened. I have got one of our fellows to address it.[48]

Goffin threatened to fire Coates as he suspected it was he who provided the student notes to Donnelly (it was not). Wisely, however, Coates left the UWS for a job at another school. In a case of indiscriminate laudation, Goffin gave him an excellent testimonial for this position – although Coates did receive a visit from his former headmaster shortly after he confessed to a vicar.[49]

In July the UWS's board of governors held its own parallel inquiry; their questions were framed around Goffin's claim that DSA exam questions were highly predictable. Other questions centred on Donnelly's tantrum.[50] For the next year the school rallied around Goffin to defend against his "persecution" by the DSA; two new societies were formed, and petitions collected, in Goffin's support. Such efforts are best explained by the fact that if his work was found to be fraudulent, then the certificates his teachers had earned might also be nullified and the school's reputation damaged.[51] They had an interest in not inquiring further into Goffin's fraud.

The letters Goffin received gave him grounds for an alternative explanation: Ledger had organized a conspiracy against him. To harm his brother-in-law's credibility, Goffin loudly and repeatedly accused Ledger of fathering a child out of wedlock (leaving unmentioned his sister's name). Although he dismissed the notes as a forgery, Goffin and the other teachers searched for the source of the leaked notes but never found out it was Jones. In January 1879 Waterlow received a mysterious letter from someone apologizing for having sent the DSA "false" information; it was composed in the style of a ransom note, with the letters and words clipped from books and newspapers and glued to a sheet of paper. Word of the letter got out, and many chose to believe that it exonerated Goffin.[52]

Waterlow convened a Select Parliamentary Committee to investigate the case, which sat in late July and early August of 1879. Waterlow represented the interests of the UWS; the DSA's interests were represented by Lord George Hamilton, then Lord President of the Committee of the Privy Council on Education; the chair was none other than Robert Lowe. Goffin's claim of a conspiracy meant that witnesses testified alone, under oath, with evidence released afterward.[53] Donnelly, Coates, and Ledger testified, with Ledger presenting crib sheets from 1872 and 1873, and a letter from Goffin warning of a DSA inspector's visit. During his day-long testimony Goffin emphasized a conspiracy, complaining about Donnelly and arguing that the authenticity of the notes could not be proven. But then an intrepid DSA clerk checked the handwriting on the exams against the handwriting on the student notes: Robert Jones was brought forward, and he reluctantly conceded that the notes were his.[54] The committee, including Waterlow, deemed that the suspension of Goffin's certificates had been justified, and in its unanimous report concluded that fraud had been "reduced to a system, and almost elevated to the dignity of an art."[55] Yet a month later, on 16 September 1879, Donnelly told his friend T.H. Huxley that "Goffin is not yet dead."[56] For Goffin remained headmaster at UWS. Waterlow had reverted to his original position, arguing that Goffin had been falsely charged. His board now announced that it would ignore the findings of the committee because the inquiry had not been conducted openly (because of Goffin's charge of a conspiracy); the board further challenged the DSA to prove its charges in open court by suing Goffin for fraud. But the DSA could not do this because Goffin received his money from the UWS board, not the DSA. Goffin then sued Donnelly for libel. The court ruled against Goffin two years later,[57] but later critics suspected Goffin's lawsuit was never intended to win anything; it was undertaken simply to prevent the DSA from discussing the case while it was before the courts.[58] After the court ruling, the final stage of the Goffin Affair played out on the floor of the House of Commons. First stood Lord George Hamilton for the DSA, who outlined the case and blamed Goffin. Then Waterlow defended Goffin, leading to charges that Goffin had bewitched him. No MP rehearsed Goffin's claims of a conspiracy; instead, his defenders made a more subtle and potent argument that gets to the heart of the problem faced by any exam system aiming for both maximum transparency and predictability. For the exams to be open, they had to be rigid, making their rules easily gamed. This format had led Goffin and other teachers to become "crammers."

In other venues Goffin had presented tabular analyses showing the frequency of certain kinds of questions on exams, such as the percentage compositions of certain chemicals or the graphic formulas of compounds and elements. By closely studying such manuals, Goffin and other successful teachers could predict what questions would be on the exam and have students memorize the answers. While such tactics were unfortunate, Goffin admitted, they were necessary for student success.[59]

Goffin did not answer why the same chemicals appeared in the notes as on the exam, and often in the same order. He obscured the issue by raising the spectre of being forced to cram. This was enough for numerous MPs, and especially Waterlow, who needed some kind of explanation to use. Back in the House, MPs sympathetically recounted how external exams had forced crammers and other coaches to study examiners' habits "with the skills of a private detective." Waterlow thought that Goffin was simply an extremely good crammer, probabilities about the Derby be damned. Indeed, he turned the matter on its head, arguing that the Select Committee's inquiry should have been public because it cast doubt upon the entire system of payment by results. It was a bad system that, by "practically offering a bribe to the masters of large schools" to almost double their salaries, enticed tutors to coach elite boys and ignore the rest of them.[60]

The Goffin Affair led to changes in when the printed exams were mailed from South Kensington: now, they were timed to arrive at the very last mail delivery before the exam was written. In 1884, for instance, only three of 5,000 mail deliveries were delayed,[61] which the department must have decided was an acceptable figure. Meanwhile, the DSA created new posts for inspectors personally responsible for the custody of the exams after they had arrived in the mail.[62] As for Goffin, he remained headmaster. Salaries of his allies tended to go up, while opponents – such as those refusing to sign one of the petitions in support – found that their teaching environment became more difficult.[63] Goffin did lose all of his DSA certificates and thus the right to earn money from that department, but Waterlow remarked that this made science teaching at the UWS more effective. All students received attention, not simply those who could perform best on exams. And the UWS continued to grow, with a lengthening waiting list – by 1881, there were 690 students at a school built for 600.[64]

This chapter is not a narrative that takes Goffin's charge of a conspiracy seriously and declines to judge. The balance of evidence strongly indicates that Goffin was guilty of defrauding the DSA, and his students, and that this fraud began in the early 1860s when he taught at Exton. It may have been prompted by his financial troubles, then grew more elaborate as his cynicism grew. Although he often fell under suspicion, because he was never significantly penalized (and often praised by officers of that same department), he likely judged the potential rewards to be worth it, and systematically expanded his system to include other teachers. Yet paradoxically, Goffin also seems to have been an exceptional teacher. His remarks on how he taught science suggest that he was passionate about the subject and much loved by his students.[65] It possible to be both an excellent teacher and someone who systematically and fraudulently manipulates the exams that are supposed to test this teaching.

What about other exams taken by Goffin's students? For they did not simply write DSA exams. One Oxford Local exam held in 1880, with a pass rate of 28 per cent, saw all seven of Goffin's students passing; in the next year, although the pass rate declined to 22 per cent, all six of his students still passed.[66] It is unclear if they succeeded because he was an excellent teacher or because of some other form of fraud. The last mention of Goffin I have found in archives are letters dated February 1893 from the Cambridge Local Examination Syndicate: its examiners had discovered widespread copying by UWS students on one of the chemistry papers, and warned Goffin to look into this. But because they did not consider the evidence conclusive, they decided to be lenient.[67] No more discussion of the matter can be found. Goffin remained headmaster of the United Westminster Schools until his retirement in 1902.

Goffin's method survives him, however, being independently rediscovered wherever decision-makers take the seemingly objective data generated by standardized tests and use it to override the judgment of teachers. In 2010, teachers at forty-four Atlanta-area schools were found to have been secretly opening standardized exams in advance, then drilling their students on the questions. The exams were part of a larger drive to measure and improve school performance – and before the fraud was discovered, the district's rising scores gained it national acclaim for its apparent reform. The irony is that the schools *were* improving, although more gradually than the tests showed – due to the expertise of the district's committed teachers. Some of the cheaters had become frustrated by the Procrusteanism of the standardized tests, which they saw as getting in the way of their real job. Others used the test scores to intimidate teachers reluctant to participate, telling them that if their students' scores didn't improve, they'd lose their job. Together, these two currents led to a larger culture of cheating in the district.[68] The Goffin Technique is one repertoire in this culture. Since cheating is only witnessed when it's unsuccessful, and since we have seen how institutions have a vested interest in not investigating themselves very thoroughly, there are probably many more cases of such misconduct than we'd like to think.

9 Economies, Remoralized: Examinations as Technologies of Inclusion

We say... 'State clearly what attainments you consider necessary for a medical practitioner; fix your standard where you please, but define it plainly...subject us ultimately to exactly the ordinary examinations and tests, and, if we fail to acquit ourselves as well as your average students, reject us; if, on the contrary, in spite of all difficulties, we reach your standard, and fulfil all your requirements, the question of "mental equality" is practically settled'...I appeal, not to the chivalry, but to the justice of the medical profession...

– Sophia Jex-Blake, "Women as Practitioners of Midwifery," 63

It is Cambridge University, 1890. Trinity College has become a Muslim college; Magdalene is now run by Cossacks; Queen's College is the domain of Australian Aborigines. Women have been admitted, and the institution responsible for advanced "FEMALE EDUCATION" is the "Cawfett Institute." The student gaining first place on the Classics Tripos that year – that is, its Senior Classic – is Miss Zuleika Spooni from Istanbul, coached by the "Hadji Shille-Tô" (Figure 5).[1]

Obviously this state of affairs did not come to pass: this vision of Cambridge in 1890 was a student's satirical projection, actually penned in 1870, of what the university might look like in twenty years' time after the just-instituted "Abolition of Tests." The "Abolition of Tests" did not end exams like the Tripos, but instead removed the test for adherence to the tenets of Anglicanism. The policy had prevented any non-Anglican from taking a degree. Now a Cambridge degree was, in principle, available to any male. With such a change, *The Moslem in Cambridge* magazine predicted that given "the gigantic strides that Feminine Intellect has made," Muslim women like Ms Spooni would be able to compete with everyone else.[2]

Figure 5. "Zuleika Spooni, Senior Classic, 1890," in Hadji Seivad [Gerald Stanley Davies], *The Moslem in Cambridge* (Cambridge: Metcalfe and Sons, 1890 [1870]), Figure 2.

As a piece of student satire, *The Moslem in Cambridge*'s humour is exuberant, inelegant, and usually offensive. Yet the author, Gerald Stanley Davies, who in 1870 styled himself "Hadji Seivad," did put his finger on something that the recent secularization of Cambridge would lead to: a greater "Cosmopolitan character" of the university.[3] He might have known that about fifty miles from the university, in Hitcham, a new institution, Girton College, that aspired to become a Cambridge college for women had just been established, and this probably would soon lead to women at the university. This was possibly the institution that the "Cawfett Institute" was referring to: the name was an anagram of Fawcett, referring to both Henry and Millicent Garrett Fawcett. Millicent Garrett Fawcett was an early feminist reformer who would go onto found Newnham College in 1875, while her husband Henry was the blind postmaster, former Cambridge professor of political economy, Liberal MP, and hero to secularists.

While religious tests were abolished, formal tests for knowledge remained just as important at Cambridge and other educational institutions throughout the British Isles. What Gerald Davies/Seivad could not have predicted in 1870 is that in the same year as his fictitious Cosmopolitan Cambridge of 1890, a woman *would* place first on the most prestigious examination in England: the Cambridge Mathematical Tripos. Phillippa Fawcett, the daughter of Millicent Garrett and Henry, wrote the exam unofficially, but under conditions made as similar as possible to those under which the male candidates had taken the Tripos. As a result, Fawcett became identified in the public mind as the de facto senior wrangler, a celebrated, well-known, and very stubborn new fact that came to be frequently deployed in various battles to attain greater rights for women.

This chapter looks at how exams were used to claim that groups previously deemed incommensurable – that is, qualitatively different – were in fact worthy of close and formal comparison. These claims of commensurability were often made in favour of those either set off to one side in a "separate sphere" or marginalized in various ways – women, non-Anglicans, non-whites, and the poor. To adopt Wendy Espeland and Mitchell Stevens's words, since exams helped to establish formal parity, they acted as a "technology of inclusion."[4]

Exams were used in three escalating commitments: as *surveys*, as *wedges*, and as *trials*. Exams used as *surveys* were like social scientific devices that gathered information about existing and newly founded schools for girls, and about female abilities more generally. Exams as surveys were also used to prove what many intuitively believed already, that female intellects were equivalent to male ones. Once these explicit proofs were established, exam successes could be used as *wedges* with which to pry open other schools to women. And finally, standardized exams became *trials* that very publicly demonstrated that since women's intellectual abilities were equivalent to men's, they were morally deserving of equitable treatment. After all, by passing exams, women had shown the merit that exams were supposed to indirectly test for. When that fair treatment was not forthcoming, the institution could be shamed.

All three types of examinations depended upon the standardization that had been so carefully crafted into examination infrastructures: not only the rules governing conduct, but even the content of the exams. The very work that secured uniformity on exams could be turned against institutions to support the claim that women ought to be judged by the same standards as men.

Chapter 5 explained how meaningful comparisons depend on the abstraction of some feature possessed by all members of a set – despite their diversity in manifold other ways – and how that feature made candidates commensurate. In the case of Victorian exams, the given feature was knowledge of a subject like Euclid, Classics, or animal physiology. Standardized exams offer emancipatory possibilities precisely because of this point. To choose to abstract one particular feature of something for comparison means that one thereby *ignores* all other features as extraneous to that comparison. In choosing to compare the physical length of a sheet of paper, stretch of road, and height of a tree, one chooses to ignore other manifold features of these entities: their density; whether they're alive or not; their colour. In choosing to compare how two candidates conjugate French verbs, one ignores, or tries to ignore, other features these candidates may possess. Such features may be banal characteristics such as height or shoe size. But they may also be characteristics that in other contexts are used to exclude or marginalize their possessor, such as gender, skin colour, age, or ethnicity.

The point about uniformity helps explain why Emily Davies insisted that her female Girton College students take exams about Euclid and other topics deemed out of date by many at the time, a decision which numerous scholars have either been unsympathetic to or mystified by. Davies's great "fanaticism" that her women students write the same exams as men, under the same conditions, followed her "principle of a 'fair field & no favour' which would generally be understood as excluding competition on dissimilar conditions."[5] Davies wished to turn the very rigidity and uniformity of exams to the advantage of women. She recognized the trap that a separate standard for women would create: a feminized standard would quickly be deemed inferior to men. By insisting upon the same, entrenched standard – even if it was antiquated – some moral claim to parity could be established.

Surveys

Widespread enthusiasm for exams had begun to wane by the early 1870s. Yet one group was keen to extend them: reformers of female schooling. Thus contemporaries noted the "remarkable" fact that although the 1867–8 Taunton Commission on Education condemned exams' overextension to boys, it recommended that exams continue to be used with girls, possibly in greater numbers.[6] The pioneering mathematics teacher Sophie Bryant praised examinations, specifically competitive ones; in her view nothing was more valuable than "finding one's level," even if the "process is not always pleasant."[7] Others who became famous as educational reformers celebrated exams too, and even payment by results. In 1874 Sophia Jex-Blake, who was campaigning for a woman's right to become a medical doctor, argued that since the principle of payment by results was becoming "daily more and more supreme," it should also be applied to medical schooling. Jex-Blake called for all candidates to write exactly the same examination, allowing all those who passed to be registered – "uniformity being required in the results alone." Jex-Blake also celebrated the competition that the exams fostered.[8] Even by 1889, amid Auberon Herbert's petition against overexamination, there remained great support for the use of exams in female schooling.[9]

In the mid-century, the few schools for girls and women that did exist – for brevity these will be called "female schools" – suffered from a double bind. On the one hand many imagined such places to have different purposes than male schools. The principle of separate spheres was followed: girls were to be instructed for the private realm, boys for the public realm. A contemporary described the assumption that "the grand use of a good education to a woman is that it improves her usefulness to somebody else ... as being companions to men, and mothers of heroes."[10] As a result there were distinctive subjects taught at female schools, especially needlework.[11] Yet on the other hand, despite these different goals, female schools *were* inevitably compared with male ones, and generally seen as inferior. It was a common belief that girls' schooling was shallower, their classes more poorly taught than boys' – a point exemplified by Dickens's 1834 satire of the "Minerva House" finishing school for young ladies, where its twenty teenage students "acquired a smattering of every thing, and a knowledge of nothing." Later criticisms were still damning: the 1867 Taunton Commission concluded that in girls' schooling there was a "want of thoroughness and foundation; want of system; slovenliness and showy superficiality; inattention to rudiments" and general shallowness.[12]

To be sure, the same criticism could be made of many contemporaneous boys' schools. Indeed, from the 1830s onward the very definitions about what constituted good schooling were being debated and contested: when it came to institutions providing formal instruction, the attributes distinguishing good ones from bad ones were still being worked out. As Joyce Pedersen has noted, much criticism of bad schools was rooted in the assumption that education ought to be individualistic and intellectualist.[13] The need for individualism and intellectual achievement was adopted by first-wave feminists as they pushed against the notion of separate spheres. This meant they emphasized sameness rather than difference between men and women, and subjects such as arithmetic, mathematics in general, modern languages, or Classics thus came to be seen as the height of meritocratic, individualistic, and intellectualist education. Reformers assumed males and females were commensurable on these subjects.

For reformers of female schools, the biggest problem they faced was the isolation of each school. Many headmistresses believed that they had insufficient information to make good decisions because they didn't know what, or how, students were being taught at other girls' schools. There was no definitive way to judge the teaching ability or qualifications of schoolmistresses. Parents, too, were thought to be in the dark about what their girls were taught.[14] For many reformers of female schools, then, exams could establish, in Pedersen's words, a "wider reference group."[15] Girls' work could be compared not only against other girls', but the individual results could be aggregated and comparisons made between various groups of girls, be they forms (classes) or schools. Publication of the results helped parents decide where to send their daughters. As a result, female students, teachers, and headmistresses at different schools could watch, judge, and imitate one another. Just as with boys' schools, girls' schools – "isolated as a rule"[16] – were now part of a larger community, drawn by the "net of examinations" into communication and commensuration with one another.[17]

Thus when the first external exams emerged with the College of Preceptors, a few reformers of female schools saw them as an ideal mode of comparison. Unfortunately most of the other reformers were also intensely concerned with class and social status, and so looked down on the College of Preceptors's exams. When Oxford and Cambridge instituted Local exams, and when London's Matriculation exams started being written outside London, many female schools saw these as preferable.[18] At the North London Collegiate School for Girls, its headmistress Frances Buss preferred these exams to inspections, which in her view provided only a general report of a school's fitness. We already noted how Buss thought inspectors (who would be male) were unsuited to judge girls' schools, partly because they were unfamiliar with the topics taught specifically to girls (such as needlework), and when they did inspect girls in subjects universally taught, they would be biased in favour of boys. As a result, Buss favoured exams that used identification numbers, or even first initials, rather than full names, so that "a French exercise, or an answer to a question in geography, may be judged by the same rules, whether written by a girl or a boy."[19]

Below, we will see how the first Locals came to be written by girls in 1863. After this, exams began to be written by more students, and reactivity began to occur: schooling for girls came to be associated with preparation for examination. By 1879, one parents' directory of girls' schools announced that since exam results were the only impartial way to ascertain if the girls were being properly taught, it would list only schools whose students took Local or other external exams.[20] Such a decision created a dilemma for headmistresses who sought to avoid these examinations. One dissenter was Frances Martin (who memorably wrote of her wish to "put a gigantic extinguisher over the existing & so called colleges for ladies")[21] – she was "very much afraid" of the examination system because she saw it as fostering vanity and competition among women. As an educator, she saw it as her calling to root out such traits.[22] Yet consciously opting out of such exams would marginalize one's own school. It is a dilemma facing any school that refuses to participate in a popular school-ranking system, even when dissenters rightly see that system as unfair or flawed.[23]

Wedges

Once commensurability in a certain field was established – the belief, for instance, that it was unproblematic to compare the geography exam results of female and male candidates – exams could be used to press further claims. The belief that exams tested not only knowledge and skill but, indirectly, desirable moral traits as well was seized upon by reformers of female schools as a means of showcasing the worthiness of female students. Moreover, the work that had gone into creating an infrastructure of examinations strengthened the belief that if a woman passed an exam previously written only by men, then that woman was entitled to exactly the same treatment as men who passed that same exam.

In this way, exams acted as "wedges" for women to pry open admittance to educational institutions. One example is how exams were used by reformers to press for women's admittance to University of London degrees. We have seen how the University of London was explicitly intended to be inclusive in its initial 1827 Charter allowing non-Anglicans to take its degrees – its language directing it to ignore "all distinctions of mere class and denomination." Despite this, it did not ignore gender, shown when Jessie Meriton White's 1856 application to become a candidate for the MD degree was rejected. More petitions to participate followed. An 1862 request by Elizabeth Garrett to take the Matriculation exam lost in the senate by only a single vote. In that same year another petition organized by Emily Davies asked that women be allowed to obtain London degrees: this petition resulted in a tied vote in the senate, defeated only by the vice chancellor's casting vote.[24] In 1866 women were given the right to earn London's "Certificates of Proficiency" rather than degrees, and the first women wrote special exams for these certificates in 1869.[25] The certificate exams differed from degree exams: they were written at a different time of year, and had some substantive differences. Women were asked geometry questions out of only the first book of Euclid's *Elements*, not the first four, but there were more language questions.

Separate exams led to two kinds of complaints. One focused on the reputation of the certificate exams: because they were held separately and written by women, certificate exams would always be deemed inferior to men's, even if the questions were more difficult.[26] Women were faced with the same double-bind problem – on one hand they were kept in a realm distinct from men, while on the other hand people would end up informally comparing them with men anyway and find them wanting. The second complaint was more prosaic, yet possibly more effective at leading to change: two separate exams meant far more work for the examiners. It meant having to create a new set of questions, administer the exam separately, then mark the completed scripts. Some London examiners asked for salary increases because of the extra work these new "special" exams created.[27] In a similar vein, the creation of separate classes for women was quietly opposed by such teachers as J.E. Cairnes: he refused to repeat his lectures, and so combined men and women into a single class.[28]

In 1874, London's senate received petitions from several women's organizations asking that women be allowed to take the men's exams. Each petition – the wording was similar or identical, indicating a coordinated campaign – focused on the inclusive wording of London's original Charter, and the exam-centred nature of its Revised Charter.[29] Since the purpose of the university was to educate all classes and denominations of Her Majesty's subjects, and since exam success was the only criterion for a BA, why did the gender of the examination candidate matter? This time the appeal to inclusiveness and uniformity, accompanied by behind-the-scenes complaints by overworked examiners, worked: in 1875 the senate declared there was no reason for there to be a difference between the women's general certificate exam and the Matriculation exam. In 1876 the certificate exam was declared "identical in all respects with that for the ordinary matriculation of men."[30] Obviously this progression was not a smooth one, and was fraught with infighting, resistance, and periodic reversals. There was a petition by numerous male London graduates, with medics being the most numerous. It is also important to note that forces external to the university played a large role in the change: the government passed a law allowing women to earn diplomas from British medical licensing bodies, giving them the right to take exactly the same qualifying exams as men. By 1878 the separate certificate exam for women was discarded, and sixty-eight women wrote the unsegregated Matriculation exam. Fifty-one passed, and by 1880 four women had gained their BAs after writing the final exam.[31]

Trials

Exams also became trials. Once the commensurability of women and men on some subject was established – making them part of a single, larger, reference group – trials were used to formally demonstrate merit and parity. The morality of exams was also played up, too. Campaigns for equal treatment demonstrated what sociologist Charles Tilly denotes with the acronym WUNC: worthiness (sober demeanour, talent, hard work), unity (individual examinees representing all women), numbers (shown by petitions), and commitment (hard word and self-discipline, possibly even self-sacrifice, exemplified by the mental and physical stamina necessary to undergo the Mathematical Tripos).[32] Publicity about the exams was always used in order to reach large audiences of people who wanted to reform female schooling. In demanding equal treatment, the reformers turned the efforts to establish examination uniformity and impartiality against those very institutions.

A case in which exams evolved from a survey, to a wedge, and then into a trial is shown in the famous campaign by women to be admitted to the Cambridge Mathematical Tripos, culminating in Philippa Fawcett's 1890 placement above the senior wrangler, demonstrating women's intellectual power as equivalent to men's. Yet the campaign actually began in 1863, and not on the Tripos but on the Cambridge Local exam. Its central figure was not Fawcett but Emily Davies, a person not always depicted sympathetically by historians of women's schooling. Her above-mentioned "fanaticism" that women write exactly the same exams as men, under the same conditions, has been described as elitist, inflexible, and limited or dismissed as sign of a "rather insecure personality."[33] Since women's education had been so patchy and inferior, scholars have argued that it was unrealistic of her to expect women to suddenly compete in Britain's top intellectual trial, not to mention subject her students to a curriculum that was increasingly being decried as antiquated.[34]

Davies's apparent "inflexibility" and "exaggerated respect" for Classics and mathematics are better explained when curricular matters are set aside and we look instead at standards and examinations' "uniformity." As we have seen, *examinations*, not instruction or lecture attendance, were seen as most important in Victorian schooling because of their ability to set common standards. For Davies there was a need for "certifying competent knowledge": "To make the discovery of individual incompetence may be wholesomely humbling or stimulating, as the case may be, but no one is the better for being told, on mere arbitrary authority, that he belongs to a weak and incapable class." Exams were important because all of the efforts to secure uniformity meant that they were not arbitrary. In support Davies invoked the case of George Eliot – how the "greatest of female novelists" had to adopt a masculine pseudonym to be taken seriously as an author – in her words, to be "measured against a class standard."[35]

Davies and other reformers of female schooling adopted a strategy that they used repeatedly. First came the survey: a petition was sent to a school or examining body requesting that females be allowed to write "experimentally" the same exam as males. Even if approval was not given, copies were secured by male colleagues and the exam written anyway, always under conditions imitating, as closely as possible, those of male candidates. Then the results were publicized, particularly successes. Female successes in those exams were not only demonstrations of women's intellectual parity with men – they also showed their candidates' moral worthiness, since passing exams required industriousness and perseverance. Reformers would then petition to allow female candidates to officially write those same exams: the wedge. If permission was given, and the exams were written, the female successes would be again publicized.

Thus on 23 October 1863 a Davies-organized petition was sent to the Cambridge Local Examination Syndicate asking if a group of London girls could be examined with the same questions that boys answered. The syndicate approved, on the condition that the grading and reporting of the exams be arranged privately between the committee and examiners.[36] Between 14–19 December of that year, eighty-three girls showed up at the London rooms of the Society of British Artists for the "experiment." The organizing committee – differing from other such committees simply by being made up of respectable ladies rather than gentlemen – ensured that the exam was written in conditions identical to those experienced by the boys. To mark the girls' exams, the committee even recruited most of the same examiners who had marked the boys' exams.[37] This first "experiment" was merely a survey: a way to ascertain how, and where, teaching at girls' schools might improve. One area of weak teaching was shown by the arithmetic results: although one examiner had privately told Davies that the "standard for a pass is not really high," thirty-four of the forty senior candidates failed the preliminary maths exam.[38]

I have not found a single mention of some innate female inferiority in mathematics as an explanation for these poor results. Instead, it was realized that the poor results were largely due to the girls' lack of familiarity with writing exams – the same explanation given five years before as to why the boys writing their first Oxford Local exams had done poorly on the arithmetic paper. Many of the girls gave wrong answers, but they had also not yet mastered the repertoire of answering an arithmetic question in a format satisfactory to an examiner. Despite these poor results, then, it was confidently predicted that with better preparation, the girls' arithmetic scores would soon "come up to the University standard" – that is, up to the boys' scores. This prediction was borne out in the following year, with only three failures among the girls. Davies's group declared their "experiment" a success.[39]

The exam results now became a wedge. In January 1864 – only one month after the exams had been written – Davies's group again petitioned the Cambridge Local Examination Syndicate, but with a new request: that girls be allowed to formally take its Locals. Girls' teachers needed a "definite aim and standard" to guide them, argued the petition. Internal exams by themselves would be less effective – compared with the well-known and impartial Cambridge Locals, internal exams were characterized as mostly useless because no one knew "how far the standard may have been lowered to meet the incompetency of the candidates."[40] The exams had done much to publicize stubborn facts that challenged the assumption that women's intellects were qualitatively different from men's, despite various efforts to establish, biologically or otherwise, innate difference. Indeed, one editorial in *The Nonconformist* claimed that girls' exam successes showed the irrelevance of the "relative weight" of male and female brains.[41]

Nonetheless, the Cambridge senate was more resistant to Davies's second petition, with opposition often grounded in concern for the girls' mental and physical health: there was a danger of female "overwork" and "exhaustion." This objection emerged not from empirical study or logic, but out of a desire to restore incommensurability due to gender difference. While Davies and other female reformers were claiming commensurability, and that it was thus fair to compare girls with boys in realms such as mathematics, her opponents claimed incommensurability – qualitative difference rooted in physical difference between females and males. Such incommensurability made examining them unfair: after all, didn't everyone know that men had stronger constitutions than women? At the same time, this so-called concern allowed opponents to explain away apparent anomalies in which women scored as high as (or higher than) men: such individual women must have freakishly strong constitutions. This meant they were qualitatively different not only from males, but from other women too – implying they were unrepresentative of women as a whole. This "no true Scotsman" form of rationalization persisted for the next thirty years.[42]

Returning to the matter of Davies's petition that girls be allowed to formally write the Cambridge Local exams: the matter went before the senate, and in a close vote, permission was granted for girls to take the Locals in 1865. There were attempts to make it difficult to compare girls' results with boys' – exam results were kept separate by gender, and only the students' identification numbers would appear in the final results. Yet although these ploys made it difficult to compare individual performances, it was not impossible. Frances Buss's own copy of the 1875 Cambridge Local exam results contains handwritten names of the North London Collegiate School's twelve candidates next to their numbers, showing that she retroactively identified them by their marks or place. In that same year the school finally had enough money to offer Sophie Bryant a post at the school to teach mathematics and German – Buss having discovered Bryant, also through the Cambridge Local exam, nine years earlier, as the only girl who got a first-class result on the mathematics paper.[43]

A further victory for reformers came with the Taunton Commission, struck in 1865, as the third inquiry to study the condition of schooling in the British Isles. As we have seen in earlier chapters, while the Newcastle Commission had studied government-funded national schools and the Clarendon studied elite public schools, the Taunton was intended to study the rest. Christopher Stray and Gillian Sutherland have nicely described the Taunton mandate as covering "every other school that had pretensions to be more than an elementary school."[44] The intention was for the commission to study boys' grammar schools, but reformers pointed out that the formal language did not rule out female schools, and so the commissioners agreed to include them in the inquiry. There, Davies testified that girls' successes on external exams made it hard for opponents to argue that girls' participation in higher education "was not in accordance with the fitness of things."[45]

In 1869 Davies founded Girton College, which eventually moved to the city of Cambridge in 1873. Although unaffiliated with the university, Girton – along with Newnham College – became one of Cambridge University's two "shadow" women's colleges. At Girton, Davies repeated precisely the same strategy as she had done with the Locals: she enlisted sympathetic male examiners to administer the same Cambridge exams that male Cambridge students wrote, and Girton students took them under exactly the conditions as the male students did. That Cambridge's curriculum was widely seen as "antiquated" – a focus on Euclid's *Elements* and Greek in particular – was regrettable, but a sacrifice Davies was willing to make in her desire for a "fair field & no favour." This single-mindedness estranged allies. Henry Sidgwick, who helped establish the other women's college at Newnham, saw Davies's desire for uniformity as too rigid; J.R. Seeley stopped teaching at Girton in protest.[46] Otherwise friendly schoolmistresses recommended their students go to Newnham rather than Girton to avoid the fate of male Cambridge students, "bound in its lifeless mechanism."[47]

One cannot help but notice in some histories a tendency to celebrate Sidgwick and others who wanted to reform the curriculum as "dynamic, forward-looking individuals,"[48] and conversely a criticism of Girton's strategy as "elitist."[49] To be sure, the frustration of the Newnham group with the "lifeless mechanism" of the broader Cambridge curriculum is understandable. But Davies was driven not by elitism but by an insistence on common standards for both men and women; precisely *what* was taught was less relevant. The examination, as an indirect marker of talent and morality, was what mattered. Indeed, following Macaulay's quip, had the Cambridge standard been fluency in Cherokee, Davies would doubtless have trained the Girton students to be examined in that language. But it was Greek, and Euclid, and other subjects, that formed the entrenched standard, and so that was the measure that would be used.

Davies's interest in examinations is best illustrated by her interest in the Cambridge Mathematical Tripos. Nowhere in Britain was the association between exam performance, intellectual capability, and impartial testing made clearer than on this Tripos. Stretched out over several brutal days, by the 1880s it consisted of papers on topics as varied as compound interest, astronomy, number theory, Euclidean geometry, trigonometry, conic sections, and hydrodynamics. The maximum possible score was around 30,000, and the person scoring highest – usually getting about half that mark – was called the "senior wrangler." Candidates receiving the next highest marks were ranked second and third wrangler, and so on. Mathematics training was supposed to teach correct reasoning, and the fame and rigour of the Maths Tripos would make a woman's success in it impossible to ignore.[50]

Although Cambridge did have other Tripos subjects in which women were informally succeeding, such as Classics, the Maths Tripos was deemed the ideal place to demonstrate intellectual parity with men. Davies declared that the Maths Tripos was to be informally written by as many Girton students as possible, in conditions as close as possible to the formal examination taken by men. One student, Sara Burstall, she dissuaded from studying history because Burstall showed some mathematical ability. Female students also honed test-taking skills through intercollegiate exams called the "Mays" – tests given to students in their first and second years, often by the lecturers.[51] While not well known, these exams were of great help in convincing new university women, often "industrious but despondent," that their work was rising to the same level as men's.[52]

There was little success in the 1870s. But in 1880, forty-one women passed the Maths Tripos, and Girton's Charlotte Angas Scott obtained a score equivalent to the male eighth wrangler's. The editors of the *Educational Times* argued that Scott's result gave women a moral claim to be formally admitted to Cambridge. Success on exams, even on informal ones, was a victory in a public trial. For Girton students could now say,

> 'We no longer ask you to let us prove our capacity for your boasted degree; we have proved it. We have shown that we can not only pass with your poll men – the students of whom you have least reason to be proud – but also with your best. We have won first-class places; we have gained your honours. Now, therefore, grant us your degrees.' We hardly see how the claim can be refused.[53]

Such pressure induced Cambridge University authorities to publish Scott's name, eighth in sequence, in its famous Senate House rankings. Later that year some 8,000 people signed a petition requesting that rather than having to depend on "the courtesy of individual examiners," women be formally allowed to take the Mathematical Tripos and other Cambridge examinations. Scott's high performance was cited in the petition.[54] The request was granted. By 1881 the first civil service posts were opened up to female clerks through competitive examination: it was in the Post Office, a move pushed by Henry Fawcett, by this point the Postmaster General.[55]

It is not stated, but as with the dispute over whether women ought to write a separate London certificate exam of their own instead of just taking the ordinary Matriculation, one wonders if a quiet factor in the decision had to do with administrative resources and efficiency. Again, running informal exams for women was complicated: permission to sit Tripos exams and have them marked had to be negotiated for each candidate, and results for individual students then communicated to their college. More female candidates wanting to take the various Tripos exams entailed more sets of negotiations and individual letters.[56] In short, it was not simply *fairer* to mark women's papers in the same pile as men's – it was also simpler, and less expensive.

Nine years later, Newnham student Philippa Fawcett famously placed 13 per cent "above the senior wrangler" in the 1890 Maths Tripos, and although she was not officially recognized as the senior wrangler, she was seen as the de facto one. Various British newspapers and even the *New York Times* announced her victory.[57] One of Frances Buss's students recalled the unforgettable moment. After reading aloud the news in a morning assembly, Buss recalled to the assembled NLCS students why Fawcett's victory mattered to her. In 1865 she and Emily Davies had been asked at the Taunton Commission whether they thought women could learn mathematics: yes, said Buss, even though at the time her school had no pupils taking it. Was she sure that women could learn mathematics? Yes, she was sure they could, Buss answered. Then she "almost shouted" to the assembled girls, "'Today those gentlemen have their answer,' and more quietly, 'I wonder how many of them are remembering, as I am remembering their question to me twenty-five years ago, and my answer!'"[58] Public demonstrations like Fawcett's undermined arguments about innate female intellectual inferiority to males.

Sousveillance?

While the account is moving, it would be simplistic to end on a triumphal note. Female examination successes like Fawcett's were often rationalized away as anomalies; at Newnham, there were attempts to pre-empt this objection by pointing to Fawcett's normality. Claire Jones suggests that women's successes on the Mathematical Tripos, rather than raising women's intellectual reputation, instead devalued that exam, or even helped to feminize mathematics over the long run, diminishing its reputation as the master discipline. And Rita McWilliams-Tullberg points out that before the Tripos victories, women in higher education were generally treated as amusing curiosities. But after results like Scott's and Fawcett's, women started to be resented as a threat.[59] The establishment of the fact that women's and men's mathematics abilities – and their intellects more generally – were commensurable promised greater competition for men in the future.

It is also important to note the other nasty little ways in which women's participation in higher education was restricted: having to use separate entrances, being prohibited from chemistry labs, or being kept in second-rate classrooms. Such barriers caused female subcultures to emerge.[60] The story of women using exams to demonstrate commensurability can only complement accounts of these restrictions: while exams may have helped "pull" girls and women into new realms of formal education, there was often little corresponding cultural "push" to actually get them into that system.[61] Any sort of equality of schooling seems to require both forces. Cambridge, for instance, did not allow women to take its degrees for another fifty years after Fawcett's Tripos victory. In discussing this long rejection, Gillian Sutherland argues that changing a culture takes more than just changing rules – it requires a shift in expectations and attitudes.[62] This chapter argues that changing beliefs in which traits *ought* to be commensurable is one specific way in which such expectations and attitudes can shift.

And sometimes development happens not when rules are changed but when people insist that existing rules be followed. The Cambridge campaign demonstrated this, and, indeed, provided a strategy for female reformers elsewhere: women gained admittance to the University of Toronto by also using the wedge tactic, unofficially writing exams that were identical to the official entrance exams that only men wrote.[63] The increasingly automatic connection made between formal exams and formal entitlements (if one passes an exam, then one gets a credential, scholarship, or position), and the insistence that the rules be followed in all cases (a candidate's identity was irrelevant, all that mattered was performance) made it possible for women to increasingly expect and demand to be treated the same as men, and downplay the differences between them. At Cambridge many women *had* demonstrated their ability on exams, which epitomized impartiality for Victorians – yet the university refused to officially recognize the link between exam success and entitlement. Such a contradiction was *why* Cambridge's continued rejection of women for degrees could be painted as discriminatory and shameful. If exams demonstrated not only intellectual ability but also desirable moral traits like competitiveness, self-discipline, and industry, then the university's foot-dragging was morally indefensible because it was self-contradictory.

Although many Victorians were sceptical that exams were supposed to discover virtuous people as well as assess knowledge, reformers of female education took this claim to be true and *acted* accordingly. This belief helped give their campaign its power. The Cambridge example illustrates how standardization sometimes helps weaker groups advance their interests: it is because of the way in which uniformity and rules get established. To standardize, rules and explicit criteria are set out and enforced, often by powerful institutional leaders. In so doing, these elites themselves often implicitly agree to be governed by the rules, subscribing to the morals behind those rules. If they don't, this discrepancy can be pointed out. Appeals to fairness and justice can be made, and the leaders shamed. Such shame can sometimes lead to institutional changes.[64]

And with this point we can return at long last to Michel Foucault, whose insights on disciplinarity and surveillance have often been deployed by scholars to explain how examinations worked, with a general picture of candidates surveilled in an educational panopticon by remote authorities. The women's exam successes show the opposite of what many neo-Foucauldians have argued. Instead of surveillance – scrutiny from above – there was a form of *sousveillance,* or scrutiny from below. For examiners, too, became constrained by the uniformity built into the new examination infrastructure. The drive to standardize sometimes meant their actions were monitored as much as examinees' were. Or the examiners saw themselves as having less and less freedom to manoeuvre. We have seen their repeated complaints about their loss of discretion to an increasingly mechanized system of exams – but sometimes this mechanized system was governed by explicit and public standards.[65] One case in which common standards constrained an examiner's discretion occurred in 1882, when Mary Scharlieb won the gold medal in obstetrics at the University of London. One male examiner in the subject privately grieved to the registrar: "To think that I have lived to give the Scholarship in my own subject to a woman! But then, it was an excellent paper and it can't be helped."[66] His hands were tied by rules and standards that he felt he had no choice but to follow.

Hence the women's exam successes had an additional effect beyond energizing their camp: they also *de-energized* any would-be marginalizers, instilling in them a sense of resignation. It is hard for individuals to oppose a group's struggle when they think their opposition is ultimately futile. It is in such an invisible form that standardization can sometimes constrain not only romantics, the poor, or the less powerful – it can also constrain would-be oppressors.

Conclusion

There was a retired sea captain who lived on a remote part of the island of Zanzibar and, as captains are wont to do, he had introduced a ceremonial raising and lowering of the flag each day, and also fired a cannon at noon. A visiting friend enquired how the captain knew the exact time, to which he replied that he reset his chronometer according to the time displayed by the clocks in the window of the jeweller in the local town. After he had left the captain, the friend was in the town and thought that he would pursue the matter further and enquired how the jeweller set his clocks. "Oh, that is simple," he replied, "there is a retired sea captain on the other side of the island who always fires a cannon at noon and I use it to set my clocks."

– Brian William Petley, *The Fundamental Physical Constants and the Frontiers of Measurement*, 8–9

In keeping Foucault at a little distance, perhaps this book has overcorrected. When this project began long ago in Ian Hacking's class, as a paper on how labels can be made to "stick" to people, it was extremely heavy on Foucault. The work rehearsed numerous points about "discipline" and "surveillance": how during the nineteenth century governments accumulated documents on people to make them legible, so they could act more effectively upon their populations. Hacking, who had just been elected to Foucault's old post at the Collège de France, found the paper simplistic. "How about documents *of* people?" Foucault himself, of course, would have agreed with this point: regarding power, individuals are conduits, not possessors: they "both submit to and exercise" power.[1] Although records can help labels stick to people, records can also give people the capability to act in different ways. They can help or hinder; they can imprison, but also set free. So too is this the case with exams, as devices that generate these records and documents. Examinations have dual, or multiple, natures.

The other French hybrid philosopher who has done much to inform the book – while also being kept at arm's length – is Bruno Latour. His work on how abstractions become mobile inscriptions has been central to this work. Flattening and simplifying complexity into pieces of paper that can themselves be reshuffled and recombined is key to being able to understand – and possibly exercise power over – that complexity. The representations generated by exams are "immutable mobiles," or, to adopt another of Latour's images, examinations are akin to the grids that scientists string across the Amazon forest floor, tools with which those scientists can orient themselves and answer a few questions about the soil's composition.[2] Such tools radically simplify a complicated situation, allowing it to be represented with abstractions. At the same time, however, if one relies too heavily on Latour's insights, there is a danger of forgetting that ultimately it is *people* who have to continuously work to ensure that the resulting abstractions remain stable.[3]

Paradoxes lurk at the heart of standardized examinations, as they do in any system that is supposed to provide public standards. We saw how exams can constrain but also emancipate. At an institutional level, exams can provide radically simplified, thin descriptions of complicated states of mind, knowledge, or organizations. Although these descriptions can be impoverished, decontextualized abstractions, such information is necessary for non-expert public servants to exercise democratic control in the public interest of accountability. Exams can neutrally measure achievement, but they can also set out targets for improvement. While publicity and openness can lend greater trustworthiness to an exam, these same qualities can also make it more rigid and therefore vulnerable to being gamed or cheated on.

Examinations also have paradoxical effects on how an individual sees oneself. If someone succeeds or fails on an examination, they usually see this as a matter of their own personal responsibility, "internalizing" standards that they might never have consciously chosen.[4] And while it is often hoped that standardized tests reward talent and knowledge, thereby encouraging the upward social mobility of an underprivileged yet deserving candidate, they often instead reward those who have already benefited from hidden cultural codes unavailable to those less fortunate. In an even more pathological way, such exams may give those privileged candidates reasons to attribute their success to their own "merit," giving moral warrant to a status they did not themselves attain individually.

Such paradoxes emerge because it is people and their performances that are being measured. Commenting on the 1890 petition against over-examination, Frederick Pollock (Trinity College Fellow, Cambridge, 1868) said that blaming standardized testing for schooling problems was like blaming a barometer for low air pressure.[5] But as with Edwin Chadwick's claim that examinations saved the labour of verification when it came to people, Pollock made a category error. Air pressure doesn't alter its behaviour based upon what a barometer says about it. What makes examinations so complex as tools is that the people they measure themselves respond to measurement. It is human looping, re-activity, reflexivity, which is the engine that powers these paradoxes. Such reflexivity means that even the Amazon forest floor will never be as complex as the developing mind of a single person, let alone a group of such minds. Soil scientists are careful not to mistake the map they create for the territory. Exams are similar tools whose representations we must also be careful not to mistake for the territory. The danger comes from mistaking the abstraction for the complex system itself – the exam results for the mind. With apologies to René Magritte, when it comes to examinations, "ceci n'est pas un esprit."

On this topic of mistaking an abstraction for the thing being studied, we can conclude with one final paradox: how the examination infra-structure, painstakingly constructed by the combined efforts of many people, came to reinforce individualistic conceptions of the mind. The efforts of numerous people to build a collective infrastructure worked so well that the resulting level grade came to be taken for granted. What emerged was a belief in the value of individualized competition, merit, and effort, as well as a reinforcement of the belief that cognition occurs entirely inside the head. Systems intended to test and reward the acqui-sition of skill and knowledge – often in forms as open and egalitarian as were possible for the time – came to reaffirm a belief in a single hierar-chy of innate mental ability, by helping to instil in people's minds what a single general intelligence might look like.

We have arrived at one way in which aptitude testing emerged out of achievement tests. By 1867 various systems had been in place for at least a decade that took candidates seen as commensurable, asked them different questions on paper, then marked their answers with numer-ical abstractions. Scores for parts of the examination were aggregated into a single number representing a candidate's entire performance on the examination, an abstraction purporting to stand for what each candidate knew. For instance, candidates wishing to become army officers had to write Sandhurst Military Academy's entrance exams. Its requirements were numerically clear: out of a total of 14,400 marks, a candidate had to get a minimum of 1,500 marks; on the mathematics section (out of 3,600), the candidate had to get a minimum of 400; on the English language section (out of 1,200), a candidate had to get 200. Other topics, each out of 1,200, were Classics, geography and history, French, German, experimental science, natural science, and drawing.[6]

In the following year, the table of Sandhurst candidates' results was analysed by someone who wished to take Adolphe Quetelet's relatively new notion of a "normal" curve and apply it to mental ability. Such a curve is now popularly known as a "Bell" curve. Quetelet had taken physical measurements such as the chest circumferences of soldiers to show that a normal curve existed in physical characteristics. This statistician analyzing the Sandhurst table, Francis Galton, argued that other physical features such as size of brains, heads, weight of grey matter, or number of brain fibres should also follow Quetelet's curve. The implication was that there was an average mental aptitude, and a "law that governs deviations from all true averages." Instead of being a horizontal normal range of chest circumferences between broad and thin, however, this would be a "vertical" normal curve of mental capacity between genius and stupidity, with most minds clustered around an average.

But how might Galton move beyond mere physical measurement to a better assessment of actual mental aptitude? One place was the table of candidates for the 1868 Sandhurst entrance exams, which constituted one public and accepted standard of achievement (Figure 6). The top candidate received between 5,800 and 6,500 marks, and although the lowest candidates' scores ruined the symmetry, Galton argued that the results of the Sandhurst exams followed Quetelet's normal curve. Nor did Galton have only the Sandhurst table to work with: an examiner for the 1868 Mathematical Tripos lent him his unofficial, confidential marking sheet as well. This allowed Galton to compare the 1868 senior wrangler – with 9,422 marks out of 17,000 – with the lowest-placed wooden spoon, who received only 309. The senior wrangler had over thirty times the marks.

Galton made two inferential leaps: first, that the senior wrangler had thirty-two times the ability of the wooden spoon, and second, that this ability was not simply in mathematics but also revealed general mental capacity. Since Galton thought it obvious that people who did not attend Cambridge must have an even lower level of attainment than the wooden spoon, he drew some highly unfavourable conclusions about the "average mental grasp" of even a well-educated audience.[7] If everyone in Britain had to take some ideal standardized test, he thought those results would also follow a normal curve.

Galton's conclusions and leaps have been well studied elsewhere.[8] What is relevant for our discussion here is that Galton could not have

Number of marks obtained by the Candidates.	Number of Candidates who obtained those marks.			
	A. According to fact.		B. According to theory.	
6,500 and above	0	73	0	72
5,800 to 6,500	1		1	
5,100 to 5,800	3		5	
4,400 to 5,100	6		8	
3,700 to 4,400	11		13	
3,000 to 3,700	22		16	
2,300 to 3,000	22		16	
1,600 to 2,300	8		13	
1,100 to 1,600	Either did not venture to compete, or were plucked.		8	
400 to 1,100			5	
below 400			1	

Figure 6. Statistical table of Sandhurst entry examination results, from Francis Galton, *Hereditary Genius: An Enquiry into Its Laws and Consequences* (London: Macmillan, 1869), 33.

made numerical comparisons of mental aptitude without there being a pre-existing infrastructure that took commensurable properties and compared them on a linear scale. The data on intelligence was not sitting out in nature, there for the picking – instead, that evidence had been pre-shaped by those systems. I suspect that the creation of standardized forms of mass assessment contributed to Galton's notion of an innate, normal form of mental ability – "general intelligence." The work done to build examination infrastructure had been so successful that he didn't notice it.

Galton's quest for a single scale of mental ability exemplified what Gerd Gigerenzer calls the "tools-to-theories heuristic," in which the tools of research come to define researchers' very conception of the phenomenon they are studying. The case Gigerenzer uses is the spread of inferential statistics in experimental psychology between 1940 and 1955. He argues that when these sophisticated statistical tools became widespread in experimental psychology as ways to obtain evidence, many psychologists started to argue that a human mind reasoned like an "intuitive statistician." This circular reasoning emerged because the methods themselves changed what kind of mental phenomena could be reported, what kinds of explanations it was possible to use, and even the kind of evidence that was being produced in the psychology laboratory.[9] In an analogous way, I suspect that as mass examining and the linear ranking of candidates' results became widespread, people like Galton began to see mental aptitude as a single, static, and commensurate property that could be discovered by mass examination and then ranked in a linear way. After all, Galton was a "Cambridge man" who had himself taken the Tripos, so he took for granted that high test results like the senior wrangler's were clear examples of extraordinary mental ability, just as others took for granted that the best evidence for teaching ability and efficiency was the exam results of their students. Like the Zanzibar paradox in this conclusion's epigraph, there was a kind of circularity at work.

To bring up the matter of circularity is not to claim that senior wranglers *did not* possess extraordinary mental ability, or that it is invalid to ascertain teaching ability on the basis of students' test results. In presenting the tools-to-theories heuristic, Gigerenzer's intention is not to denounce its apparent circularity. He wishes only to remind us not to forget about alternative explanations: there are always different possible maps that can be generated of a particular territory. There are other explanations for how people reason, other than as intuitive statisticians; there are forms of intelligence other than mathematical ability; there are ways to assess knowledge and skill other than by standardized testing.

So we can end this book by noting alternatives to Galton's on how one might better measure mental ability. One 1882 author (possibly William Chambers) half-jokingly suggested that "brain-power" would be better assessed by skill at games like chess rather than at mathematics. An excellent test of mental ability was the three-move chess problem. But an even better test was the card game of whist, an ancestor of today's game of bridge. The author went so far as to claim that while a "first-rate mathematician may on other subjects be stupid; a first-class whist-player is rarely if ever stupid on original matters requiring judgment." For unlike mathematics, or even chess, whist could better gauge intelligence because the values of the cards shifted during the play of the game.[10] Since such changing values meant many more possible combinations to choose from, this made whist less mechanical, and better rewarded a player's awareness of the changing local conditions of a game.

One further reason why whist (or bridge) is interesting as a test of mental ability is that since such games are played with partners, and against other pairs, each player is not an isolated individual but part of a group. The game arises from the continuous and sensitive readjustment of each player to the others' actions, actions that are themselves looping, reflexive reactions to the other players. This form of "tuning"[11] is another way in which the Zanzibar anecdote is appropriate. And so this book can close on its own half-joking note. Perhaps skill at bridge – or some other social game in which the variables change during play – might supplement today's exams, as additional devices with which to identify not only mental ability, but also the exercise of judgment and sensitivity to nuance. Such traits would also seem to be important in any society that aspires to be a meritocracy.

Appendix A: Important Dates

1730	University of Cambridge's Senate House Honours Examination instituted
1731	Competitive examinations begin at Trinity College, Dublin
1800	System of public examinations begun at Oxford University (modified 1807)
1822	Senate House Honours Examination becomes Mathematical Tripos
1846	University of London Matriculation Examinations first written overseas, in Sydney
1846	Queen's Scholarship exam for pupil-teachers becomes first proctored examination
1851	First girls examined alongside boys by the College of Preceptors
1852	Oxford Commission recommends more examinations
1853	Northcote-Trevelyan Report recommends competitive examinations for candidates to Indian Civil Service
1855	Department of Science and Art exams held at Irish Industrial Museum, Dublin
1856	Society of Arts's first examination, London
1856	Jessie Meriton White petitions to be examined to enter University of London's MD program
1857	Society of Arts' first proctored examination, London and Huddersfield
1857	Oxford Associate-in-Arts examination held in Exeter

1857	New Charter of University of London permits degrees to be earned only by examination results
1860	DSA exams using "payment on results" begin
1861	Newcastle Commission on "Popular" [elementary] schooling issues report
1862	Revised Code links elementary school funding to examination results
1863	Cambridge Local Examinations written in Port-of-Spain, Trinidad
1863	First girls write Cambridge Local Examinations
1864	London's Matriculation Examination written in Port Royal, Mauritius
1867	Taunton Commission on "middle class" schooling for girls and boys issues report
1868	Elizabeth Garrett Anderson petitions to be examined in order to enter University of London's MD program
1869	Women write the first Certificate of Proficiency exams at the University of London
1869	Girton College founded by Emily Davies
1872	Devonshire Commission on Scientific Instruction issues first report
1878	R.E.H. Goffin caught opening DSA exams in advance
1878	Women's Certificate of Proficiency exams at University of London discarded; women write Matriculation exams with men
1880	Charlotte Angas Scott places equivalent to eighth wrangler on Mathematical Tripos
1885	Government of Madras [Chennai] institutes payment by results for science exams based on DSA model
1890	Philippa Fawcett scores "above the senior wrangler" on Mathematical Tripos
1890	Auberon Herbert organizes petition against "over-examination"

Appendix B: Biographical List

Many of these bios are drawn from the *Oxford Dictionary of National Biography*, edited by H.G.C. Matthew and Brian Harrison, 61 vols (Oxford: Oxford University Press in association with the British Academy, 2004).

Abney, William (1843–1920). Royal Engineer and inspector of science schools in the DSA, working for Donnelly. Fellow of the Royal Society in 1876 in chemistry.

Acland, Thomas Dyke (1809–1898). Devon politician. Graduated BA, double first, Christ Church College, Oxford, 1835. Friends with W.E. Gladstone. In 1857, with Frederick Temple, Acland helped organize the first Oxford Local Examinations in Exeter.

Anderson, Elizabeth Garrett (1836–1917). Medical reformer. By gaining the licence of Society of Apothecaries in 1865, she became the first UK-trained woman to go on the British Medical Register. In 1868 Anderson petitioned to enter the University of London's MD program. In 1870 she gained a University of Paris MD, and she supported the London School of Medicine for Women.

Armstrong, Henry Edward (1848–1937). Chemist and advocate of the "heuristic" (discovery) method of science teaching. He studied at the Royal College of Chemistry with Hofmann then Frankland, getting his PhD in Leipzig in 1870.

Arnold, Edward (?–1887c). School inspector. Brother of Matthew Arnold. After Rugby and Balliol College, Oxford, he became an All Soul's Fellow in 1851. He was appointed assistant inspector in 1854 and a full inspector in 1863 for Devon, Cornwall, and the Scilly Isles.

Arnold, Matthew (1822–1888). School inspector, poet. After Rugby School, he gained a scholarship to Balliol College in 1841. His 1847 position as secretary to the Whig politician Lord Landsdowne helped him become a school inspector. Arnold toured France in 1859 for the

Newcastle Commission and in 1868 for Taunton Commission. In 1870 he became a senior school inspector.

Beale, Dorothea (1831–1906). School reformer, headmistress. After receiving Queen's College diplomas certifying her ability to teach, she became head teacher at its preparatory school in 1854. She became principal of Cheltenham Ladies' College in 1858.

Besant, Water (1836–1901). Eighteenth wrangler, Christ's College, 1859. In 1860 he became senior professor at the Royal College, Mauritius, where with Frederick Guthrie he organized a revolt against its rector, bringing in the University of London Matriculation Examination as an independent assessment of academic standards.

Booth, James (1806–1878). Leading ideologue of national school examinations. After graduating from Trinity College Dublin in 1833, Booth wrote fellowship exams four times, unsuccessfully. He became principal of Bristol College in 1840, then vice principal of Liverpool Collegiate Institution. Booth proposed a UK-wide examination scheme in 1846. As chairman of the Society of Arts, he ran and expanded its examination scheme in 1856 and 1857, alienating Harry Chester in the process.

Briggs, William (1861–1932). Created University Correspondence College (UCC). Briggs won the Akroyd Scholarship to the Yorkshire College of Science, Leeds, in 1874, while remaining mathematics teacher at his own secondary school. After being appointed science professor at St Benedict's College in Scotland in 1883, he remotely coached Leeds friends for the University of London mathematics examinations. With his wife, Ada (daughter of a postman, knowledgeable about mail pricing), he expanded his remote teaching practice into the UCC.

Bryant, Sophie (1850–1922). Examinations pioneer, headmistress. Bryant moved to England from Ireland when thirteen; in 1866 she gained a scholarship to Bedford College and was only girl that year with a first-class result on the mathematics portion of the Cambridge Local Examinations. She was hired by Frances Buss to teach mathematics at the North London Collegiate School (NLCS); in 1884 she became the first woman to receive London's Doctorate in Science. Bryant became headmistress of the NLCS in 1895 and was the first woman elected to the senate of the University of London.

Buss, Frances Mary (1827–1894). Schooling reformer, headmistress, North London School for Girls. Buss established the school in 1850. As its first headmistress, she strongly pushed for competitive examination of girls. In 1869 she became the first female Fellow of the College of Preceptors.

Carpenter, William (1813–1885). University administrator. He received an MD from Edinburgh in 1839 and became Fullerian professor in physiology, Royal Institution, in 1844. Carpenter served as registrar of the University of London from 1856 to 1879.

Chadwick, Edwin (1800–1890). Philosophical radical. Private secretary of Jeremy Bentham, 1830–32. Chadwick is most famous for his 1842 "Report on the Sanitary Condition of Labouring Population of Great Britain." In 1828, he proposed examinations modelled on French *concours*. Friends with Henry Cole.

Chester, Harry (1806–1868). Civil servant with the Committee of the Privy Council on Education. Chester helped to establish its 1846 teacher examinations. He became chairman of the Society of Arts in 1853 and designed its first examination scheme in 1855, which was a failure. Friends with Henry Cole; bitter enemy of James Booth.

Cole, Sir Henry (1808–1882). Bureaucratic reformer, empire builder. Friends with philosophical radicals, Cole pushed for uniformity and reform of public records (1835–8), a uniform penny post, and uniform narrow railway gauges. A major organizer and publicist of the Great Exhibition of 1851, he used leftover funds to purchase land in South Kensington for an arts and science institution. As a believer in the "piecework" principle of payment, Cole devised, tested, and refined at least three schemes tying examination results to direct or indirect payment.

Dasent, George Webbe (1817–1896). Examiner for Civil Service. He earned a BA from Magdalen Hall, Oxford, in 1840 with no distinction, but he became an expert in Scandinavian literature and thence gained a professorship of English literature and modern history at King's College, London, where he examined for the Civil Service.

Davies, Emily (1830–1921). Schooling reformer, founder of Girton College, Cambridge. Davies canvassed with Elizabeth Garrett to open University of London degrees to women. To get women into university examinations, she formed a committee that persuaded the Cambridge Examination Syndicate to do so in 1863. Davies successfully lobbied the Taunton Commission to include girls in an investigation of the status of middle-class education. She founded Girton, the first women's college at Cambridge, in 1869.

Deighton, Horace (1831–1913). Twenty-first wrangler, Queen's College, Cambridge, 1854; teacher then headmaster of Queen's Collegiate School 1860–72, Port of Spain, Trinidad. Authored a textbook of Euclid's *Elements*.

Donnelly, John (1834–1902). Brilliant administrator. Royal Engineer and Crimea veteran, Donnelly returned to London in 1856 as commander of the South Kensington detachment. Hired by Henry Cole to reorganize the DSA, he became inspector for science in 1859, figuring out how to implement Cole's ideas. He worked to institute the City and Guilds Institute for Technical Education Exams. Friends with T.H. Huxley.

Fawcett, Henry (1833–1884). Liberal politician. He was taught at Queenwood College by Hirst and became seventh wrangler (Trinity Hall), Cambridge, in 1856. Fawcett was blinded in accident in 1858. He was elected professor of political economy, Cambridge, in 1863; elected Liberal MP, 1865; married Millicent Garrett, 1867; father of Philippa Fawcett, 1868. Became Postmaster General, 1880.

Fawcett, Philippa (1868–1948). First female senior wrangler (unofficial), Newnham College, Cambridge, 1890. Daughter of Millicent Garrett and Henry Fawcett. After University College London courses in pure and applied mathematics, she gained a scholarship to Newnham in 1887. On 7 June 1890, she scored "above the senior wrangler" on the Cambridge Mathematical Tripos, creating a worldwide sensation.

Fitch, Joshua (1824–1903). School inspector. University of London BA, 1850; appointed school inspector, 1863. Fitch acted as assistant commissioner for the Taunton Commission. In 1878, as a member of the University of London senate, he helped formulate the New Charter putting female and male students on equal terms. He was a close ally of Emily Davies.

Frankland, Edward (1825–1899). DSA chemistry examiner. Frankland passed Playfair's exam at the Putney College of Engineering and became his assistant. He earned a PhD in Marburg in 1849 and taught at Queenwood College with Tyndall and Hirst. When he replaced August Hofmann at the Royal College of Chemistry, he became the chemistry examiner for the DSA exams.

Gladstone, William (1809–1898). Liberal politician. Gladstone turned a famed "double-first" examination result from Christ Church, Oxford (1831) into a political career, moving from conservativism to liberalism, where he held cabinet positions such as Chancellor of the Exchequer and Prime Minister. He generally favoured financial retrenchment. For Gladstone, examination success was self-evidently a marker of talent and gentlemanly status. During the 1852–4 Oxford reform question, he worked closely with Jowett and Temple; on the Civil Service examination question, he worked closely with Northcote and Trevelyan.

Goffin, Robert Edwin Hemblington (1836–1922). Infamous examination cheater. Goffin was a gifted but financially troubled teacher who saw the earning potential of DSA's payment-by-results scheme, taking its very first teachers' exams 1859. He was also among the first to take Huxley's training courses for teachers in the summer of 1872. Goffin became headmaster of United Westminster School in 1874. In

1878 he was shown to have been systematically opening DSA exams in advance and drilling students on the questions and answers, leading to the "Goffin Affair."

Grote, George (1794–1871). Radical politician and architect of the University of London's "New Charter" of 1856. A successful yet reluctant banker, Grote wrote a twelve-volume history of ancient Greece, emphasizing its democratic politics. He was a protégé of James Mill, and his social circle included Jeremy Bentham, John Stuart Mill, and other philosophical radicals. He was one of London University's founders, 1827–30; Radical MP, 1832–41; member of the senate of University of London, 1850–71; and Vice Chancellor, 1862–71.

Guthrie, Frederick (1833–1886). DSA physics examiner. A University of London graduate, Guthrie taught at the Royal College of Mauritius, 1861–7, where he pushed for use of the University of London Matriculation Examination. He worked with Huxley to establish summer laboratory courses for physics teachers. He taught H.G. Wells, who he thought was a weak student.

Guthrie, George (1785–1856). President of Royal College of Surgeons. Military surgeon in Wellington's Peninsular army. Guthrie became President of Royal College of Surgeons in 1833, where he orally questioned candidates. His gruff manner seems to have hidden a softer interior: "[T]hough dreaded as an examiner, he never rejected a candidate by his unsupported vote" (*Oxford Dictionary of National Biography*).

Hirst, Thomas Arthur (1830–1892). DSA mathematics examiner. Hirst followed his friend John Tyndall from Queenwood College to Marburg, where he also gained his mathematics PhD. In 1873 he became director of the Royal Naval College.

Hooker, Joseph Dalton (1817–1911). Examiner in botany, surgery. Gained his University of Glasgow MD in 1839; friend of Darwin and Huxley. Hooker became examiner in surgery for the East India Company and for the Apothecaries Company's botany medal in 1855. In 1865 he became director of Kew.

Huxley, Thomas Henry (1825–1895). Science examination ideologue. Impoverished scholarship student and University of London medallist who became a naval surgeon and naturalist. He taught natural history and physiology at School of Mines from 1854 onward. Huxley was also an examiner at the University of London from 1856 onward, and for the DSA, 1859 onward. He created the South Kensington training course on teaching biology in 1871.

Jowett, Benjamin (1817–1893). Champion of examinations. Jowett earned a first-class BA from Balliol College in 1838 and became a Balliol Fellow that same year. He was the centre of a network of friends

(Matthew Arnold, Lingen, Northcote, Temple) who went on to change various British schooling systems. Jowett coached private pupils and helped design the Indian Civil Service examinations. "Mr. Jobbles" in Trevelyan's *The Three Clerks* was based on Jowett.

Kay-Shuttleworth, James (1804–1877). Schooling reformer. He was first Assistant Secretary to the Committee of the Privy Council on Education, which administered the government grant to education. Kay-Shuttleworth opposed the Revised Code of 1862.

Lankester, Edwin (1814–1874). DSA botany examiner. A self-taught botanist, Lankester made a name as a popular author in microscopy and botany; in 1859 he became an examiner in botany for the DSA. He was a public health reformer and an advocate of teaching physiology in schools; his son Edwin Ray became one of Huxley's star pupils.

Latham, Henry (1821–1902). Cambridge examiner. Eighteenth wrangler, 1845, Trinity College, Cambridge. Latham moved to Trinity Hall, where as tutor, then master, he expanded that college substantially.

Ledger, Charles (dates unknown). Goffin's pupil, Ledger learned the "Goffin system" of opening examinations in advance and drilling for the DSA exams, writing seven in May 1871. He married Goffin's sister, Jane, in 1875, and became a teacher at the United Westminster Schools that year. Ledger fell out with Goffin in 1877.

Lingen, Ralph (1819–1905). Balliol Fellow 1841, friend of Jowett's. Lingen succeeded James Kay-Shuttleworth as Assistant Secretary of the Committee of the Privy Council on Education. Supported competitive examinations. A skilled civil servant and Lowe loyalist. When Lowe became Chancellor of the Exchequer he followed him, giving Lingen the freedom to bully other civil servants about costs.

Lodge, Oliver (1851–1940). Physicist. His interest in science was strengthened by attending John Tyndall's Royal Institution physics lectures and succeeding on Tyndall's DSA physics exams. After taking the University of London Matriculation exam in 1872, Lodge studied under Huxley, Frankland, and Guthrie at the Royal College of Science. He gained London's Doctorate in Science in 1877 and went on to promote Maxwell's theory of electromagnetism.

Lowe, Robert (1811–1892). Liberal politician. Overcame near-blindness with feats of memory and boundless self-confidence. Lowe was a successful Classics tutor at Oxford from 1834 to 1838, leader-writer for *The Times*, and a Liberal MP. As joint Secretary to the Board of Control from 1852 to 1855, Lowe opened up Indian Civil Service positions to competition, following the Northcote-Trevelyan report. As Vice President of the Education Department (1855–7, 1858–64), he tied examination results to funding in the infamous Revised Code of 1862. As Chancellor of the Exchequer, he instituted examinations across the entire Civil Service.

Lubbock, John (1834–1913). "Scientific" MP. As son of the banker-astronomer John William Lubbock, he succeeded his father in banking and science (entomology and anthropology). A friendship with Charles Darwin and his status as a Liberal MP made him a valuable conduit between Liberal power centres and scientific men eager to transform the English schooling system.

Magnus, Philip (1841–1933). Spokesman for private technical instruction. He gained a University of London BA in 1862 and BSc in 1863. As a renowned tutor for the London exams, he was drawn into the practice of schooling. He became the director of the City and Guilds of London Institute for the Advancement of Technical Education, which took over the trade and technical exams of the Society of Arts in 1879.

McConnish, Margaret (dates unknown). Pioneer. In 1871 she won the prize for best work in T.H. Huxley's first summer physiology course at South Kensington, out of thirty-nine students. She went on to become a demonstrator in Huxley's laboratory.

Mill, John Stuart (1806–1873). Philosopher. Philosophical radical, associate of George Grote, and later champion of Liberalism. Mill was a proponent of examinations as vehicles for institutional reform.

Moseley, Henry (1801–1872). Examiner for military schools. Seventh wrangler, St John's, 1826; appointed professor of Natural and Experimental Philosophy and Astronomy at King's College, London, in 1831. Moseley became an inspector of teacher training schools in 1844 and was a member of the Council of Military Education.

Norris, John Pilkington (1823–1891). School inspector. Graduated Trinity College, Cambridge, 1846; Trinity Fellow, 1848. Norris was an HMI for Staffordshire, Shropshire, and Cheshire from 1849 to 1864.

Northcote, Stafford (1818–1887). Liberal politician. Graduated Balliol 1839, becoming Gladstone's secretary. One of the secretaries of the Great Exhibition. He jointly authored the Northcote-Trevelyan report, which recommended competitive examinations to join the Indian Civil Service. "Sir Warwick Westend" in Trollope's *The Three Clerks* was based on Northcote.

Perry, John (1850–1920). Engineering professor. A DSA night school student and a foundry apprentice during the day, Perry was "discovered" after winning prizes on DSA examinations, securing a Whitworth scholarship to study engineering full time in 1868. He became a professor of civil engineering at the University of Tokyo.

Playfair, Lyon (1818–1898). Chemist, Liberal MP, examination ideologue. Gained PhD in Liebig's chemistry laboratory, invited to sit on commissions by Peel. After acting as special commissioner of the Great Exhibition, Playfair was appointed professor of chemistry at the Government School of Mines, and Secretary for science in the DSA. He was one of the first to warn of a British industrial "decline" caused by a lack of scientific knowledge among its workers. As a Liberal MP, his 1875 report supported competitive exams in the Civil Service.

Ramsay, Andrew (1814–1891). DSA geology examiner. Ramsay became chair of Geology at the Royal School of Mines in 1851, which meant he was asked to become a DSA examiner.

Roscoe, Henry (1833–1915). DSA chemistry examiner. After studying at the University of London, Roscoe gained his PhD with Robert Bunsen. In 1857 he replaced Edward Frankland as professor of chemistry at Owens College. He served as a Liberal MP from 1885 to 1895.

Sayce, Archibald Henry (1845–1933). Examinations critic. A college tutor at Oxford in comparative philology and Assyriology, Sayce claimed that examinations were giving the UK "an adopted Chinese culture."

Scott, Charlotte Angas (1858–1931). Mathematical pioneer. In 1876 Scott went to Girton College on a Goldsmith's Company scholarship, and in 1880 gained a score equivalent to the eighth wrangler's on the Mathematical Tripos. In 1885 she obtained a London DSc and became a professor of mathematics at Bryn Mawr College, USA.

Stephen, Leslie (1832–1904). Critic of schooling and examinations. Twentieth wrangler, 1854, Trinity Hall; Trinity Hall Fellow, 1854.

Tagore, Satyendranath (1842–1923). Examinations pioneer. Tagore sailed to London from Kolkata to write the 1863 Indian Civil Service exams, and was the first native-born inhabitant of South Asia to succeed. He returned to India, starting as a magistrate in 1864 in Ahmedabad, and ending his career as a sessions judge in Satara. An Indian nationalist and a feminist *avant la lettre*; older brother of Rabindranath.

Temple, Frederick (1821–1902). Champion of examinations. As an impoverished Balliol student, in May 1842 Temple gained a double first, then became a Balliol Fellow. Friends with Jowett, Arnold, Lingen. He worked for the Committee of the Privy Council on Education as principal of Kneller Hall, the teacher training college, and became an inspector of teacher training colleges. In 1857 he helped institute Oxford's Associate-in-Arts ("Local") Examination, adopting practices from the Society of Arts and College of Preceptors. Became headmaster of Rugby School, 1857. Member of Taunton Commission; Bishop of Exeter, 1869; Archbishop of Canterbury, 1885.

Todhunter, Isaac (1820–1884). Mathematics educator, textbook author. BA, University of London, 1842; gained its Gold Medal in Mathematics, 1844. In 1848, Todhunter became senior wrangler and Smith's Prizeman, St John's College, Cambridge. He wrote numerous bestselling mathematics textbooks, such as Euclid's *Elements*.

Trevelyan, Charles (1807–1886). Examinations ideologue. Graduated East India Company's Haileybury College at top of class, then entered Bengal Service, 1826. From 1840 to 1859 he was Assistant Secretary to the Treasury, where he supported changes such as civil service examinations. "Sir Gregory Hardlines" in Trollope's *The Three Clerks* was based on Trevelyan.

Tyndall, John (1820c–1893). DSA physics examiner. Taught at Queenwood College, Hampshire, 1847–8, 1851–3. Gained his PhD, Marburg, 1851, with Robert Bunsen. Professor of Natural Philosophy, Royal Institution, 1853.

Wallace, Alfred Russel (1823–1913). DSA physiography assistant examiner. Co-discoverer of the principle of evolution by natural selection, Wallace made extra money by marking DSA physiography examinations (though not setting the questions).

Waterlow, Sydney (1822–1906). Liberal politician. Printer and chairman of United Westminster Schools from 1873 to 1893 – in essence, Robert Goffin's employer.

Wells, Herbert George (1866–1946). Coach. After earning numerous DSA grants for his headmaster, he received a scholarship to Huxley and Guthrie's Normal School of Science, South Kensington, in 1884. In 1890 he received a University of London BSc and became a successful tutor for William Briggs's University Correspondence College.

White, Jessie Meriton (1832–1906). Pioneer. In 1856 her petition to be the first woman to write the University of London Matriculation Examination was rejected. She became a leading proponent of the unification of Italy.

Wilson, James Maurice (1836–1931). Examinations sceptic. A senior wrangler (St John's, 1859), he forgot all of his higher mathematics just after completing the Tripos. He became mathematics master at Rugby School in 1859. Wilson was sceptical of Huxley's claim that learning science would automatically hone one's skills of induction.

Wren, Walter (1834–1898). Coach. Graduated 1852 from Gonville and Caius College; afflicted by spinal injury. Founding partner of Wren and Gurney's Preparatory Establishment, where he repeatedly charged the

Public Schools with lobbying for exam standards to be lowered in order to protect their own students. Radical Liberal.

Wynne, Anne (unknown–1874). Worried parent. Edward Wynne's mother and wife of John Tyndall's first patron, George Wynne of the Royal Engineers.

Wynne, Edward (1849–unknown). Examinee. "Eddy," Anne's son. On his third try, he passed the entrance exam to Woolwich Military Academy, which trained Royal Engineers. Privately taught by Tyndall and Hirst.

Notes

Introduction

1 Repertoires are less deterministic an explanation than "culture": there are a range of repertoires to choose from in a given situation. The word can also refer to enactments continuously created and maintained. Faulkner and Becker, *Do You Know: The Jazz Repertoire in Action*, 94, 192; Tilly, *Popular Contention in Great Britain*, 41–4.
2 Bowker, "Information Mythology," 234–5.
3 Lampland and Star, "Reckoning with Standards," 17.
4 Armstrong, "Chemistry," 222.
5 Sayce, "Results of the Examination-System," 843; Elman, *Civil Examinations and Meritocracy in Late Imperial China*, 10.
6 Fyfe and Lightman, "Introduction," 11.
7 Butterworth, "The Science and Art Department 1853–1900," 502–4; Barton, *The X Club*, 303.
8 Ambirajan, "Steam Intellect and the Raj," 173; Jarrell, *Educating the Neglected Majority*, 231–3, 270.
9 Reese, *Testing Wars in the Public Schools*.
10 This would be what Heather Douglas defines as $objectivity_2$ in her "The Irreducible Complexity of Objectivity," 459; she gives eight versions of objectivity.
11 Espeland and Sauder, *Engines of Anxiety*, 4, which describes how rankings shape U.S. law school behaviour rather than simply describe their relative merits. They adopted this phrase from Donald MacKenzie's *An Engine, not a Camera*, which describes how financial models actually shape markets rather than simply describe their workings.
12 Hoskin, "The 'Awful Idea of Accountability,'" 270.
13 Foucault, *Discipline and Punish*, 189–93; Foucault, *Power/Knowledge: Selected Interviews*.

14 Porter, "Thin Description," 225; Frith, "Socialization and Rational Schooling," 80.

15 See, for instance, the point by Gareth Stedman Jones that one must not simply treat individuals as the assignees of subject positions in discursive practices (i.e., discussions) but also study how those discursive practices are changed by actions taken by those same individuals – "The Determinist Fix: Some Obstacles to the Further Development of the Linguistic Approach to History in the 1990s," 24–5.

16 Indeed, in a different context (global meteorology), Paul Edwards has suggested that the mutual shaping of human agency alongside sociotechnical systems and institutions be called "infrastructuration"; Edwards, "Meteorology as Infrastructural Globalism," 239–40. "Infrastructuration" is probably a more accurate label to describe the mutual enactment of examiners, examinees, and examination systems, but the term will probably alienate most readers, so it will not be used.

17 Booth, *How to Learn and What to Learn*.

18 Rothblatt, "Supply and Demand: The 'Two Histories' of English Education," 639.

19 Lyon Playfair to John Tyndall, 7 January 1857, Royal Institution MS JT/TYP/3/982–983.

20 Barton, *The X Club*; Becker and Clark, *Little Tools of Knowledge*; Clark, *Origins of the Research University*; Craik, *Mr Hopkins' Men*; Delve, "The College of Preceptors and the Educational Times"; Eddy, "The Interactive Notebook"; Eddy, "The Shape of Knowledge"; Elman, *Civil Examinations and Meritocracy in Late Imperial China*; Gould, "Femininity and Physical Science in Britain"; Hansen, "Rethinking Certification Theory"; Jacobs, "Girls and Examinations"; Jarrell, "Visionary or Bureaucrat?"; Jarrell, *Educating the Neglected Majority*; Jones, "Femininity and Mathematics at Cambridge"; Reese, *Testing Wars*; Schwartz, "Professions, Elites and Universities in England"; Simon, "Secondary Matters"; Simon, *Communicating Physics*; Stray, *Assessment in Education*; Stray, *Classics Transformed*; Stray, "Educational Publishing"; Stray, *Oxford Classics*; Stray, "The Shift from Oral to Written Examination"; Stray and Sutherland, "Mass Markets: Education"; Warwick, "A Mathematical World on Paper"; Warwick, "Making Sense of Maxwell's Treatise"; Warwick, *Masters of Theory*; Zeng, *Dragon Gate*.

21 Brock, "Geometry and the Universities"; Brock, "School Science Examinations"; Butterworth, "Science and Art Department"; Butterworth, "Science and Art Department Examinations"; Dyhouse, *No Distinction of Sex?*; Foden, *James Booth and the Origin of Common Examinations*; Gooday, "Precision Measurement"; Kuhn, *The Essential Tension*; Layton, *Science for*

the People; MacLeod and Moseley, *Days of Judgement*; McWilliams-Tullberg, *Women at Cambridge*; Pedersen, *Reform of Girls' Secondary and Higher Education*; Rouse Ball, *A History of the Study of Mathematics at Cambridge*; Roach, *Public Examinations in England*; Rothblatt, "Student Sub-Culture and the Examination System"; Sadler, "The Scholarship System in England to 1890"; Sadler and Sislian, *Representative Sadleriana*.

22 Coined in Young's 1958 *Rise of the Meritocracy*, he intended the word "meritocracy" to denote not a desirable system but an undesirable one.

23 Elman, *Civil Examinations and Meritocracy in Late Imperial China*, 384n136.

24 Carson, "Differentiating a Republican Citizenry," 102–3. He further develops his argument in *The Measure of Merit*.

25 Andrew Warwick makes this counterintuitive but convincing point in "A Mathematical World on Paper."

26 That is, if in a triangle two angles are equal to each other, then the sides opposite those equal angles are also equal to each other.

27 "King Cram in India," 233–4.

28 Gladstone to Victoria, 17 February 1854, in Benson and Esher, eds., *The Letters of Queen Victoria*, 3:11.

29 Macaulay et al., "Report on the Indian Civil Service," 15.

30 Civil Service Commissioners, *Third Report*, xxviii; Lowell, *Colonial Civil Service*, 95; Moses, *Civil Service of Great Britain*, 57; Caro, *Power Broker*.

31 Booth, *How to Learn and What to Learn*; Rolleston and Turner, *Scientific Papers and Addresses*, l–li; Chadwick, "Open Competitive Examinations," 671; Northcote, "Competitive Examination for the Civil Service," 279–80.

32 Edgeworth, "Element of Chance in Competitive Examinations"; Edgeworth, "Uncertainty of Examinations"; Roach, *Public Examinations*, 283.

33 Desrosières, *History of Statistical Reasoning*, 23; Latour, *Pandora's Hope*, 42; Agar, *The Government Machine*, 1–3.

34 Espeland and Stevens, "Commensuration as a Social Process," 316.

35 Ibid.

36 See, for instance, Carson, *The Measure of Merit*.

37 This is to borrow some ideas from Ginzburg, "Morelli, Freud and Sherlock Holmes," 14; Huxley, "On the Method of Zadig."

38 Busch, *Standards*, 52–3, 59.

39 Brereton, *The Case for Examinations*, 23–4.

40 Hacking, "Looping Effects."

41 Carson, *The Measure of Merit*, 61–3.

42 Espeland and Sauder, "Rankings and Reactivity," 2–7.

43 Hacking, "Looping Effects," 368–70.

44 Hacking, *Historical Ontology*, 111.

45 "Standards of Examination in Elementary and Class Subjects."
46 Secord, "Knowledge in Transit," 665.
47 Ambirajan, "Steam Intellect and the Raj," 173; Grigg, "Scientific and Technical Education in Madras," 107.
48 Fyfe and Lightman, "Introduction," 11.
49 Lightman, *Victorian Popularizers of Science.*
50 Josep Simon has suggested historians of science pay greater attention to the history of education. *Communicating Physics*, 17.
51 Topham, "Scientific Publishing," 561.
52 White, *Thomas Huxley*, 72–5; Brock, "Spectrum of Science Patronage."
53 Wallace, *My Life*, 2:406–11.
54 Arnold and Russell, *Letters of Matthew Arnold*, 381–2.
55 Desmond, *Huxley*. The dull routines of examining are discussed in Elwick, "Economies of Scales."
56 Lightman, *Victorian Popularizers of Science*, 180, 284; Desmond, "Redefining the X Axis," 14.
57 Schaffer, "Modernity and Metrology," 71–2; Kula, *Measures and Men*, 288; Alder, "A Revolution to Measure," 62; Porter, *Trust in Numbers.*
58 Stinchcombe, *When Formality Works*, 17.
59 Fyfe and Lightman, "Introduction," 4.
60 Ibid., 9.
61 [de Morgan], "Review of *How to Learn and What to Learn*," 1531.
62 Endersby, *Imperial Nature*, 6; Reidy, *Tides of History*, 12, 92–3.
63 Foucault, *Discipline and Punish*, 189.
64 Porter, "How Science Became Technical"; Porter, "Thin Description."
65 I am grateful to Simon Schaffer for this phrase.
66 Foucault, *Discipline and Punish*, 189–92; Hoskin, "The Examination"; Hoskin and Macve, "Accounting and the Examination."

1. "The Age of Examinations"

1 Mill, qtd in *Re-Organisation of the Civil Service*, 95. Regarding physical measurements, M. Norton Wise has argued that any proposed measure of assessment must represent some valuable quality for the relevant audience so that it is not seen as "some artifact of arbitrary or interested opinion." Wise, "Introduction," 7.
2 Caplan, *Case against Education.*
3 Sutherland, "Education," 169.
4 Foucault, *Discipline and Punish*, 176; see also Hoskin and Macve, "Accounting and the Examination," 106–7.
5 Norris, *Education of the People*, 180.

6 Taunton Commission, *Minutes of Evidence* II, C. Cassal (Q10681, 10684, 10739, 10742, 10756), 169, 181, 183, 188. There are probably links between these different national visions of schooling and the centralized workings of civil law in France versus a more decentralized common law in the British Isles.
7 Taunton Commission, *Minutes of Evidence* II, C. Cassal (Q10739), 181–2.
8 Newcastle Commission, *Report*, 35.
9 Shteir, *Cultivating Women*, 228–9.
10 Davies, *Application of Funds to the Education of Girls*, 3–4.
11 This is to use one school inspector's phrase; Newcastle Commission, *Minutes of Evidence*, F. Watkins (Q1141), 155.
12 The boys from each school ended up coming into "rough contact" anyway, fighting in "tribal battles" on a nearby hill with wooden clubs and stones wrapped in scarves. Sherborne, *H.G. Wells*, 30–1.
13 Newcastle Commission, *Minutes of Evidence*, E.C. Tufnell (Q3308), 1:410–11.
14 Historian Henry Holman draws this link between monitors and pupil-teachers in *English National Education*, 106.
15 Taunton Commission, *Minutes of Evidence* I, J.P. Norris (Q425), 50.
16 Stray, *Classics Transformed*, 32, 87.
17 I thank Marsha Richmond for this extremely helpful "pulling" and "pushing" image.
18 Kafka, *The Demon of Writing*, 34.
19 [de Morgan], "Review of *How to Learn and What to Learn*," 1531.
20 Warwick, "Mathematical World on Paper"; Warwick, *Masters of Theory*; Ball, *A History of the Study of Mathematics at Cambridge*; Gascoigne, "Mathematics and Meritocracy"; Sadler, "Scholarship System in England"; Stray, "Shift from Oral to Written Examination"; Clark, *Origins of the Research University*; Craik, *Mr Hopkins' Men*.
21 Stray, "Shift from Oral to Written Examination," 36–7, 42; Clark, *Origins of the Research University*, 111.
22 Clark, *Origins of the Research University*, 113–17, 31–4. Michael Hoskin and Richard Macve argue that the "mark" – a number assigned to each examination answer, which could be added up to yield a representation of the student's overall knowledge, first appeared in the "engineering" exams pioneered by William Farish in 1792; "Accounting and the Examination," 127–8. This is disputed by Christopher Stray, who cites Rouse Ball (1889, p. 213) in pointing out that the first year in which "impression marking" was not used at all was 1836. Even this evidence does not conclusively show that 1836 was the first year in which numerical marks were assigned to individual questions. Stray, "Shift from Oral to Written Examination," 40–1.
23 Clark, *Origins of the Research University*, 131–4.
24 Porter, *Trust in Numbers*, ix.

25 Daston and Galison, *Objectivity*, 121. Daston and Galison are discussing the procedures to illustrate scientific atlases, but the definition also fits highly rigid procedures followed by tests such as the Cambridge Maths Tripos. See also Douglas, "The Irreducible Complexity of Objectivity."
26 Friedland, *University of Toronto: A History*, 22.
27 G.O. Trevelyan, *Life and Letters of Lord Macaulay*, 2:344.
28 Latham, *On the Action of Examinations*, 35.
29 Clark, *Origins of the Research University*, 131–4.
30 Warwick, *Masters of Theory*, 174.
31 Rothblatt, "Examination System in Early 19th Century Oxbridge," 84–6, 279.
32 Hooper, "College Life at Cambridge," 461–2.
33 Oxford University Commission, *Inquiry* (Report), 57–60; Sadler, "Scholarship System in England," 53–4; Foden, *James Booth*, 6–9, 12–13.
34 Oxford University Commission, *Inquiry* (Report), 57–60.
35 Stray, "Shift from Oral to Written Examination," 43–5; Oxford University Commission, *Inquiry* (Report), 60–1.
36 Trollope, *The Three Clerks*, I.233–4.
37 Sylvester, *Robert Lowe and Education*, 4, 24.
38 Oxford University Commission, *Inquiry* (Evidence Pt. 1), 12–13.
39 Danvers et al., *Old Haileybury College*, 51–3, 58–60, 228–40.
40 Agar, *The Government Machine*, 47–8.
41 Danvers et al., *Old Haileybury College*, 374.
42 Agar, *The Government Machine*, 49–50.
43 C.E. Trevelyan, *The Irish Crisis*, 201; Agar, *The Government Machine*, 51–2.
44 Trollope and Skilton, *An Autobiography*, 75.
45 Rodger, *Command of the Ocean*, 120–2, 511–12. N.A.M. Rodger claims the decision was not made by Samuel Pepys, as is commonly thought, but Charles II, who alone had the power to impose the requirement that men with very different social statuses be subject to the same test, which was highly unusual at the time. Although favouritism was shown, it was not to sons of the nobility, but to those of naval officers.
46 Hearl, "Military Examinations," 113. Wellington himself had been sent to a French military academy in Angers because he had been thought stupid.
47 Sturges and Sturges, *In the Company's Service*, 50–1.
48 Hearl, "Military Examinations," 109.
49 Taunton Commission, *Minutes of Evidence* I, H. Moseley (Q1827), 188; *The Times*, 15 January 1855, 4.
50 Hearl, "Military Examinations," 133.
51 Desmond, *Politics of Evolution*, 11, 28–30.
52 *Second Statement by the Council of the University of London*, 26–7.
53 University of London, *Papers Relating to the Origin of the University* (1857), University of London Archives ULC PC2/15, 2; University of London, *Minutes of Committees* (1847–50), 3–4.

54 R.W. Rothman to [T.H.] Huxley, 22 September 1841, University of London Archives RO 1/2/3.

55 This caution is added because Benjamin Elman claims that this was first done in China for its famous Imperial Exams; Elman, *Civil Examinations and Meritocracy in Late Imperial China*. However, as I understand it, the Chinese system was eliminative in a series of "heats" until the victor of a given province was declared – quite a different process than printing off questions in advance and sending those questions elsewhere, with the examiners then generating an abstraction of the answers (i.e., in numbers such as percentages) and ranking those abstractions. A sympathetic and highly detailed Victorian account of the procedures governing Chinese exams can be found in "A Competitive Examination Hall in China."

56 University of London, *Minutes of the Senate* (1843–9), 33–4, 45–8.

57 Rather than a department, as this would have led to conflict with the Church of England.

58 Boulger, *Candidate's Complete Instructor*, 76–7.

59 Minutes of the Committee of the Privy Council on Education, 24 September 1839, qtd in Holman, *English National Education*, 87.

60 Kay-Shuttleworth, *Four Periods of Public Education*, 330–2.

61 Boulger, *Candidate's Complete Instructor*, 76–7.

62 Tropp, *School Teachers*, 14, 18–19; Sutherland, *Elementary Education in the Nineteenth Century*, 21.

63 "Organization of the Civil Service," *Journal of the Society of Arts* 2, no. 67 275.

64 Dickens, "Hard Times, Chapter I."

65 Taunton Commission, *Minutes of Evidence* I, W.B. Carpenter (Q888), 100.

66 College of Preceptors, *Fifty Years of Progress in Education*, 3–8.

67 Delve, "College of Preceptors and the Educational Times," 52–3, 145–6. Sylvester was the unofficial second wrangler; unfortunately, given the prejudices of the day, as a Jew he was prohibited from obtaining a Cambridge degree. Wilson had been his Tripos coach.

68 Willis, "Market Forces and State Intervention in Educational Enterprise," 98; College of Preceptors, *Fifty Years of Progress in Education*, 13–14; Jacobs, "Girls Have Done Very Decidedly Better Than the Boys," 123; Delve, "College of Preceptors and the Educational Times," 168.

69 Taunton Commission, *Minutes of Evidence* I, J. Robson (Q20), 3–4; "College of Preceptors – August Distribution of Prizes and Certificates," 215–16.

70 "Discontinuance of the Educational Examinations of the Society of Arts," 255; Wood, *History of the Royal Society of Arts*, 426–7; Foden, "Technical Examinations in England," 72–3.

71 Samuelson Committee, *Report*, J.C. Buckmaster (Q8204), 412.

72 [Chadwick], "A General, Medical, and Statistical History," 551–4. Perhaps the *concours*'s most interesting feature was its use of sortition: questions

were placed in an urn and randomly selected by each candidate, who then answered viva voce.

73 Charles Edward Trevelyan to John Delane, 6 February 1854, Gladstone Papers, vol. 248, British Library Add. MS 44,333 ff. 138–49; Hart, "Genesis of the Northcote-Trevelyan Report," 69, 71. Mill's praise of Chadwick's paper is noted in Administrative Reform Association, *Unfitness of the Present Home Government*, 15.

74 John Lefevre to Charles Edward Trevelyan, 30 January 1854, Gladstone Papers, vol. 248, British Library Add. MS 44,333 ff. 109–10.

75 King, *Guide to the Civil Service Examinations*, viii–ix; Agar, *The Government Machine*, 52; Trevelyan to Delane, 6 February 1854, Gladstone Papers, vol. 248, British Library Add. MS 44,333 ff. 138–49; *Re-Organisation of the Civil Service*, 43.

76 King, *Guide to the Civil Service Examinations*, vii; Mann, "Civil Service Appointments," 303.

77 Preston-Thomas, *Work and Play of a Government Inspector*, 2.

78 Agar, *The Government Machine*; Roach, *Public Examinations*.

79 Richard Dawes to the Earl of Aberdeen, 20 February 1854, cited in Hart, "Genesis of the Northcote-Trevelyan Report," 80.

80 William Edward Gladstone to Queen Victoria, 17 February 1854, qtd in Benson and Esher, eds., *The Letters of Queen Victoria*, 3:11.

81 Qtd in Hughes, "Civil Service Reform, 1853–5," 28–9.

82 Stray, *Classics Transformed*, 38.

83 Trollope and Skilton, *An Autobiography*, 28–31. This was a posthumous criticism.

84 "Educational and Literary Summary of the Month," September 1863, 130.

85 "Hindoo Civil Servants," 375.

86 Acland, *New Oxford Examinations*, 101; Willis, "School Examinations from 1850 to 1917," 101.

87 Green, "Archbishop Frederick Temple on Meritocracy," 58, 154–5.

88 Acland, *Middle Class Education*, iii, viii, 49–50; Acland, *New Oxford Examinations*, 8, 16, 75–6, 101.

89 Fortescue, *Public Schools for the Middle Classes*, 4–7, 90; Delve, "College of Preceptors and the Educational Times," 168; "Middle Class Examinations," 226.

90 *Middle Class Education*, vol. 2, iii, v, 42, 49–50; Temple, *Middle Class Examinations*, 34.

91 Chester, "On the Society of Arts' Union of Institutes," 195–6.

92 "Oxford Associate-in-Arts Examination," 187.

93 The same pattern can be seen with the *US News and World Report* rankings of U.S. law schools. Espeland and Sauder, "Rankings and Reactivity," 18.

94 Dalgleish, "University Certificate Examinations," in *Transactions*, 274–5; Dalgleish, "University Certificate Examinations," *Educational Times*, 190–1.
95 Norris, "On the Proposed Examination of Girls," 408–9.
96 The preliminary exam had one paper requiring writing from dictation in order to test candidates' spelling, but this was done en masse, and the papers were marked later.
97 Porter, "Thin Description," 221–2.
98 Fowler, "On Examinations," 418; "Collapse of Wallahism," 540.
99 Alborn, "Quill-Driving," 46–7.
100 Clarendon Committee, *Report*, 139.
101 University of London, *Minutes of Committees* (1853–66), University of London Archives ST 3/2/6, 66–9.
102 Stockwell, "Examinations and Empire," 206–7; *Occasional Papers on University Matters*, 41; Taunton Commission, *Minutes of Evidence* I, B. Price (Q730), 81.
103 Taunton Commission, *Minutes of Evidence* I, G. Liveing (Q133), 15.
104 "Oxford Local Examinations," September 1863, 123.
105 Taunton Commission, *Minutes of Evidence* II, G. Rawlinson (Q537), 61–2; (B. Price Q658), 74; "Educational Summary of the Month," June 1880, 159.
106 Also referred to as the Queen's Royal College, later; for consistency "Queen's Collegiate School" is used, since it's the nomenclature adopted in the 1870 Keenan Report. Keenan, *Papers on the State of Education in Trinidad*.
107 Stockwell, "Examinations and Empire," 206–7; Cambridge Local Examination Syndicate, "Minutes" (1862–7), Cambridge University Library, LES 1/1.
108 Sedbergh School, *Register, 1546 to 1895*, 223.
109 Cambridge Local Examination Syndicate, "Minutes" (1862–7), Cambridge University Library, LES 1/1.
110 Gordon, *A Century of West Indian Education*, 36, 41–2, 74, 232–3.
111 Campbell, *Colony & Nation*, 26–9, 99–100; de Verteuil, *The Holy Ghost Fathers of Trinidad*, 11–13; Gordon, *A Century of West Indian Education*, 52–5, 241–2.
112 Royal College Rector, Mauritius to Chairman of College Committee, 20 April 1864, University of London Archives RC 40/20; Bruce, *Instruction in Mauritius*, 33–8; Governor Sir H. Barkly to Secretary of State for the Colonies Edward Cardwell, 29 July 1865, University of London Archives RC 40/20; Besant, *Autobiography*, 118–19.
113 William B. Carpenter to Sir Frederic Rogers, 15 July 1864, University of London Archives RO 1/2A/4; Bruce, *Instruction in Mauritius*, 11–13.
114 William B. Carpenter to Dean of Cape Town, 6 December 1866, University of London Archives RO 1/2A/6.
115 Gilchrist Educational Trust and University of London, *Conditions for Scholarships*, 1–2; Carteret-Bisson, *Our Schools and Colleges, for Boys*, 129–30.
116 Carteret-Bisson, *Our Schools and Colleges, for Boys*, 524.

117 "Examinations Letter Book 1890–1891," Cambridge Assessment Archives A/LB 1/2.
118 "Education in the British Colonies," 231–4.
119 *Cambridge Review*, 28 November 1895. The target of the poem must have enjoyed it, as a copy can be found in John Neville Keynes's "Common-Place Book," papers of Dr J.N. Keynes, Cambridge Assessment Archives, PP/JNK/1/3. Keynes was the head of the Cambridge Local Examination Syndicate.
120 Taunton Commission, *Minutes of Evidence* II, G.W. Dasent (Q13946, 13959), 517, 519.
121 Whewell, *English University Education*, 55.
122 MacLeod, "Introduction," 7.
123 G.O. Trevelyan, *Competition Wallah*.
124 "Eighth Annual Conference," 570.
125 J.M. Wilson, "On Teaching Natural Science in Schools," 269–72.
126 Earl Granville, 128 *Parliamentary Debates* L., 3rd ser. (1853), col. 38; "China and the Chinese," 536–7. There are claims that the original impetus for paper examinations in Europe came from China, particularly through accounts of missionaries such as Matteo Ricci. His reports about the use of exams in China to advance one's social position through merit were read by *philosophes* like Voltaire and Turgot; Ssu-yu Teng, "Chinese Influence on the Western Examination System." However my research could not find direct links from China to Great Britain. In a remarkable innovation to preserve impartiality, the Chinese system required that each essay be recopied by clerks before marking, thus anonymizing candidates. See Morris, "An Historian's View of Examinations," 2–3; Sadler, "Scholarship System in England," 55–8; Hudson, *Europe & China*, 325–6. A fascinating Victorian account of how a competitive exam was run in China can be found in "A Competitive Examination Hall in China," 323–4. For more general attitudes to their system, see in Alcock, "The Peking Gazette," 344–5; Martin, "Competitive Examinations in China," 75–6.
127 Rowe, *China's Last Empire*, 48. I thank Hsiang-Fu Huang for this point.
128 Roberts, *Concise History of China*, xii.
129 Playfair, *On Teaching Universities and Examining Bodies*, 2, 10; Hutchinson Almond, *Mr. Lowe's Educational Theories*, 5–6; Arnold, "The Twice-Revised Code," 358. It is Almond who called Lowe the "British Confucius."
130 "The Encouragement of Scientific Research," 486.
131 Sayce, "Results of the Examination-System," 843.
132 "College of Preceptors – March Distribution of Prizes and Certificates," 81; "Analyses of Books, Experimental Chemistry for Junior Students," 240.

133 Gilbert and Sullivan, *Iolanthe; or, the Peer and the Peri*, https://gsarchive.net/iolanthe/html/index.html, accessed 26 March 2020. I thank Allan Olley for alerting me to this point.
134 Herbert, "State Education," 53.
135 Herbert, "Sacrifice of Education to Examinations," *The Nineteenth Century*, 617.
136 Pollock, *Oxford Lectures*, 226–7.
137 Holman, *English National Education*, 158.

2. Monetizing Marks

1 Holman, *English National Education*, 141.
2 Goldman, "Victorian Social Science," 92.
3 Scott, *Seeing Like a State*.
4 Forgan and Gooday, "Constructing South Kensington," 437.
5 Butterworth, "Science and Art Department," 502–4; Barton, *The X Club*, 303. The number given by Butterworth is 2,002,529 candidates; Barton's numbers are slightly different.
6 Barton, *The X Club*, 303.
7 Donnelly, *Instruction of the Industrial Classes*, 11.
8 Layton, *Science for the People*.
9 [Lankester], "Instruction to Science Teachers at South Kensington."
10 Simon, *Communicating Physics*, 17.
11 The Committee of the Privy Council on Education was in the Privy Council as an experiment, with the Privy Council being memorably referred to as "a kind of potting shed for new administrative plans"; Sutherland, *Policy-Making in Elementary Education*, 22. Only with the 1856 move of the DSA out of the Board of Trade did this group gain the moniker "Education Department." For consistency I will refer to it as the Education Department. For a contemporary description of the Privy Council's various roles see Boulger, *Candidate's Complete Instructor*, 71.
12 Butterworth, "Science and Art Department," 46–7.
13 Sneyd-Kynnersley, *H.M.I.*, 112–13.
14 Rupke, *Richard Owen*, 68.
15 Donnelly, *Instruction of the Industrial Classes*, 10.
16 Barton, *The X Club*, 99.
17 Finer, *Life and Times of Sir Edwin Chadwick*; Hart, "Genesis of the Northcote-Trevelyan Report." For a definition of "revolution" see MacDonagh, "The Nineteenth-Century Revolution in Government."
18 Cole and Cole, *Fifty Years of Public Work*, 2:18–20.
19 Hill believed that simplifying the system would lead to mail traffic increasing inversely as the squares of the prices, a scheme that Martin

Daunton finds opaque in his *Royal Mail* (p. 8) – but that is likely just Rowland Hill, former schoolmaster of Mill Hill, leaning on a quasi-Newtonian formula for rhetorical support.

20 Daunton, *Royal Mail*, 5–6; Hill, *Post Office Reform*, 21–4.

21 Cole and Cole, *Fifty Years of Public Work*, 1:34, 47–8, 68; Hill, *Post Office Reform*, 21–4; *First Report from the Select Committee on Postage*, R. Hill (Q114), 13; Daunton, *Royal Mail*, 18. Indeed, Cole, along with Earl Granville, Stafford Northcote, and Lyon Playfair would in 1853 push for the extension of this uniform postal system to the entire British Empire; Cole, "Colonial and International Postage."

22 Williams, *Our Iron Roads*, 192–4. The advantage of the broad gauge was that it permitted more room for the locomotive engine; it was also believed to offer a more comfortable ride for the passengers.

23 Williams, *Our Iron Roads*, 194; Cole and Cole, *Fifty Years of Public Work*, 1:77–80; Bonython and Burton, *Henry Cole*, 73–4.

24 Butterworth, "Science and Art Department," 9, 16.

25 Henry Cole, "Diaries 1853"; Cole, "Diaries 1854."

26 Cole, "Diaries 1853"; Cole "Diaries 1854."

27 Cole, *Functions of the Science and Art Department*, 12.

28 Oxford University Commission, *Inquiry* (Evidence Pt. 1), 12–13.

29 Butterworth, "Science and Art Department," 189; "History of the Science and Art Department" (1883c), xxxiii.

30 Goldman, "Victorian Social Science," 92.

31 Taunton Commission, *Report*, 2:61–2.

32 Samuelson Committee, *Report*, J.F.D. Donnelly (Q76), 5.

33 Lowe, *Middle Class Education*, 8.

34 Taunton Commission, *Report*, 2:61–2.

35 Lowe, "Shall We Create a New University?," 164.

36 Cole, "Diaries 1853"; Cole, "Diaries 1854"; *Re-Organisation of the Civil Service*, 46–7, 54, 244, 375–80.

37 Cole, "Memorandum on the Corps of Royal Engineers," 15 September 1869.

38 Chadwick letter, 1 August 1854, qtd in *Papers on the Re-Organisation of the Civil Service*, 200.

39 "Piece-Work at South Kensington," 104. Although the target of the joke was Lowe – *Punch* portrayed Cole as Lowe's reluctant instrument of the policy – it equally applied to Cole and his love of piecework.

40 DSA, *Précis of the Minutes of the Science and Art Department*, 123; "History of the Science and Art Department," xxxviii; Cole, "Memorandum on the Objections Made by Sir J.P. Kay Shuttleworth," 15 May 1861.

41 Post Office, *Uniform Inland Penny Postage*, 71–2.

42 Playfair, *Industrial Instruction on the Continent*.

43 Butterworth, "Science and Art Department Examinations," 28.

44 Boulger, *Candidate's Complete Instructor*, 79.

45 Administrative Reform Association, *Appointments for Merit*, 64–5.
46 Becker, "On the Study of Science by Women," 394–5; Playfair to Tyndall, 7 January 1857, Royal Institution MS JT/TYP/3/982–983.
47 Playfair to Tyndall, 7 January 1857, Royal Institution MS JT/TYP/3/982–983.
48 Golinski, *Making Natural Knowledge*, 180–2; Porter, *Trust in Numbers*, 75, 78.
49 Qtd in Barton, "'Huxley, Lubbock, and Half a Dozen Others,'" 426–7, 432.
50 Bonython and Burton, *Henry Cole*, 185; Reeks, *Royal School of Mines*, 126–30. Donnelly's father had been stationed at Addiscombe, the East India Company's military college, so John was raised in a scholastic environment.
51 "Seventh Annual Conference," 502, 504–5.
52 Cole, "Diaries 1857"; see also Bonython and Burton, *Henry Cole*, 178.
53 Mason, "Genesis of Payment by Results," 272–3; Cole, "Diaries 1859"; Bonython and Burton, *Henry Cole*, 191.
54 Cole, "Diaries 1859"; Devonshire Commission, *Report*, xix–xx; *Register of the Associates and Old Students of the Royal College of Chemistry*, lvi–lvii; *Crockford's Scholastic Directory for 1861*, 8; Barton, *The X Club*, 304; DSA, *Directory* (1869), 6–9.
55 Cole, "Memorandum on the Corps of Royal Engineers," 15 September 1869; Horn, "The Corps of Royal Engineers," 21.
56 Horn, "The Corps of Royal Engineers and the Growth of British Technical Education," 23
57 Huxley to Hooker, 6 October 1864, qtd in Butterworth, "Science and Art Department," 221–2; Samuelson Committee, *Report*, T.H. Huxley (Q8000), 401.
58 Huxley, *Collected Essays*, 3:131–2.
59 Armytage, "J.F.D. Donnelly," 14–15.
60 Huxley, "National Education in Science and Art," *The Times*, 31 January 1887, 4; *Nature*, "The Calendar and General Directory of the Science and Art Department," 321. The letter is unsigned, but Harry Butterworth believes that Donnelly asked Huxley, the skilled propagandist, to write it: Butterworth, "Science and Art Department," 146. For Huxley's early assistance in helping to start *Nature*, see Baldwin, *Making Nature*, 26-7.
61 Ambirajan, "Steam Intellect and the Raj," 173; Grigg, "Scientific and Technical Education in Madras," 103–4.
62 Government of Madras, "Resolution of the Government of Madras [1886]," 116–17.
63 Donnelly, *Lecture on the Promotion of Science Instruction*, 8, 15.
64 Bonython and Burton, *Henry Cole*, 198; Boulger, *Candidate's Complete Instructor*, 76–7.
65 Mason, "Genesis of Payment by Results," 272–3.
66 F. Smith, *Sir James Kay-Shuttleworth*, 192.
67 Lowe, "Speech on the Revised Code," 199–200; Newcastle Commission, *Report*, 324–5.

68 Ball, *Her Majesty's Inspectorate 1839–1849*, 206–9; Newcastle Commission, *Report*, 234.
69 Committee of the Privy Council on Education, *Minutes* (1842–3), 30.
70 Newcastle Commission, *Minutes of Evidence*, F.C. Cook (Q860–4), 126–7; Newcastle Commission, *Minutes of Evidence*, F. Watkins (Q1041), 147; "Educational and Literary Summary of the Month," May 1861, 43.
71 Newcastle Commission, *Minutes of Evidence*, R. Lingen (Q426), 57.
72 Newcastle Commission, *Report*, 238.
73 Lowe, "Speech on the Revised Code," 203–4.
74 Newcastle Commission, *Report*, 273; Lowe, "Speech on the Revised Code," 202–3.
75 "Statistics of Education in England and Wales," *The Times*, 28 March 1861, 8; see also Rigg, "Establishing a Revised Code," 610.
76 Newcastle Commission, *Report*, 324–5, 334; see also Rigg, "Establishing a Revised Code," 586–7.
77 Mason, "Genesis of Payment by Results," 275–7.
78 Taunton Commission, *Minutes of Evidence* I, R. Lowe (Q6650), 642.
79 Henry Cole, "Diaries 1861"; Bonython and Burton, *Henry Cole*, 199–200; Cole, "Memorandum on the Objections Made by Sir J.P. Kay Shuttleworth," 15 May 1861.
80 Butterworth, "Science and Art Department," 90.
81 The Act gained this name, "Revised Code," because it *codified* past minutes and orders.
82 Lowe, *Middle Class Education*, 3, 8.
83 *Minute Establishing a Revised Code*, 3, 8.
84 Stray and Sutherland, "Mass Markets: Education," 369.
85 Committee of the Privy Council on Education, *Report* (1862–3), xxi–xxii; *Minute Establishing a Revised Code*, 9.
86 Arnold, "The Twice-Revised Code," 360.
87 Arnold's fame as a poet did not translate into universal respect for his work as an inspector – other educationists saw him as out of touch with contemporary developments, too trusting, and sometimes even filled with "contempt for the business." See, for instance, Browning, "Report on the System of Education," 476.
88 "The Working of the Revised Code," 367.
89 Duke, "Robert Lowe – a Reappraisal," 23.
90 Lord Sherbrooke [Lowe] to Lord Lingen, 17 March 1882, qtd in Martin, *Life and Letters of the Right Honourable Robert Lowe*, 217.
91 See Walpole, 166 *Parliamentary Debates* H.C., 3rd ser. (1862), cols. 35–7; Fitch, "Educational 'Results,'" 261.
92 Butterworth, "Science and Art Department," 91.

3. An Epistemology of the Mundane

1 Butterworth, "Science and Art Department," 195; Desmond, "Redefining the X Axis," 28. The division between "elementary" and "advanced" was created in 1865, and it is possible that this category of "elementary" led to what Huxley would go on to call elementary biology, not the Elementary Education Act of 1870.

2 DSA, *Nineteenth Report* (1872), 28; Devonshire Commission, *Report*, T.H. Huxley (Q350), 25; Donnelly, *Instruction of the Industrial Classes in Elementary Science*, 11.

3 Simon, *Communicating Physics*, 71.

4 DSA, *Directory* (1872), 129–34; compare with the 1869 directory.

5 DSA, *Directory* (1872), 101, 106, 112–3.

6 Indian Civil Service, *Selection and Training of Candidates*, 71–2; Simon, *Communicating Physics*, 182.

7 Goffin, "Mr. Goffin's Certificate."

8 Indian Civil Service, *Selection and Training of Candidates*, 71–2.

9 Select Committee, *Mr. Goffin's Certificate*, Appendix 4, 228.

10 Select Committee, *Mr. Goffin's Certificate*, Appendix 4, 229.

11 Ibid.

12 Practical exams began in 1878, and only for the inorganic and organic chemistry exams.

13 Uzzell, "The Teaching of Chemistry," 129.

14 Devonshire Commission, *Report*, xxiii–xxvi; Devonshire Commission, Appendix IV, 7–8.

15 Taunton Commission, *Minutes of Evidence* I, G.D. Liveing (Q241), 27–31; John Tyndall, "Journals," 1855–72, Journal VIa, 17 April 1857, 927.

16 DSA, *Eighteenth Report*, 49; Devonshire Commission, *Report*, 17; Devonshire Commission, Appendix IV, 7–8.

17 T.H. Huxley to Michael Foster, 15 April 1872, Huxley Papers, Imperial College Archives, 4.38; T.H. Huxley to Michael Foster, 5 May 1874, 4.81.

18 Hints of fraud can be found in William Miller's questioning of Henry Cole on 14 June 1870 for the Devonshire Commission; Miller himself alluded to this problem regarding the proofs at the University of London. Devonshire Commission, *Report*, W. Miller questioning H. Cole, Q218–19), 15–16. As a professor of chemistry at King's College and as a University of London senate member, Miller may have been referring to proofs of medical exams being leaked beforehand.

19 Devonshire Commission, *Report*, T.H. Huxley (Q252), 19; Devonshire Commission, *Report*, H. Cole (Q216, 218), 16.

20 Sneyd-Kynnersley, *H.M.I.*, 112–13.

21 Select Committee, *Mr. Goffin's Certificate*, J.F.D. Donnelly (Q138, Q150-2), 18–19; J.F.D. Donnelly and W. Abney (Q373–6), 40–1.
22 Warwick, *Masters of Theory*, 18. His section (pp. 18–26) is titled "Entering the House of Theory." Baron Rayleigh's name was John William Strutt.
23 Located in Surrey, to the south-southwest of London.
24 Select Committee, *Mr. Goffin's Certificate*, C. Ledger (Q1484–8), 89–90.
25 Donnelly, *Instruction of the Industrial Classes in Elementary Science*, 5–7; Select Committee, *Mr. Goffin's Certificate*, C. Ledger (Q1491), 90; *Papers and Correspondence between Science and Art Dept. and Committee of Class No. 3150*, 36.
26 Huxley, *Lessons in Elementary Physiology*, 91.
27 Alcock, *Questions on Huxley's Lessons*, 15–19.
28 Select Committee, *Mr. Goffin's Certificate*, Appendix 4, 228–31.
29 Select Committee, *Mr. Goffin's Certificate*, Appendix 6, 266. Frustratingly, despite the DSA reports' extraordinary detail on matters such as payments made and rules to be followed, documentation on the dates and locations of specific exams could not be located. The local committee, which administered the region's exam, should not be confused with the Parliamentary Select Committee, which later ran an inquiry into Goffin's examination practices.
30 Select Committee, *Mr. Goffin's Certificate*, Appendix 5, 241.
31 DSA, *Eighteenth Report* (1871), x–xi; DSA, *Minutes*, 333; DSA, *Twenty-First Report* (1874), 43–4.
32 DSA, *Twenty-First Report* (1874), 13–14.
33 Trollope, *The Three Clerks*, 1:239–40.
34 Examinee [pseud.], "Notes in the Senate House," 79.
35 Select Committee, *Mr. Goffin's Certificate*, Appendix 4, 228–31.
36 Select Committee, *Mr. Goffin's Certificate*, C. Ledger (Q1488), 90.
37 DSA, *Twenty-First Report* (1874), 13–14.
38 Devonshire Commission, *Report*, Appendix IV, 7–8; Devonshire Commission, *Report*, T.H. Huxley (Q252, 256, 265, 267), 19.
39 Devonshire Commission, *Report*, Appendices III and IV, 6–7; DSA, *Ninteenth Report* (1872), 43.
40 Devonshire Commission, *Report*, Appendix III, 6–7; Devonshire Commission, *Report*, T.H. Huxley (Q252, 256, 265, 267), 19–20.
41 Butterworth, "Science and Art Department," 198; Devonshire Commission, *Report*, Appendix III, 6–7.
42 DSA, *Twenty-First Report* (1874), 33, 145. Goffin's gold medallist was Ethelbert Dowlen, a twenty-four-year-old clerk.
43 Select Committee, *Mr. Goffin's Certificate*, Appendix 6, 265–6.
44 Eve, "On Marking," 12–13.

45 Instead, Ledger's completed script was found in Parliamentary Committee evidence. Just why it was located there will be revisited in a later chapter.
46 Foden, *James Booth*, 11–12.
47 Select Committee, *Mr. Goffin's Certificate*, Appendix 4, 228–31.
48 See, for instance, Leys, *From Sympathy to Reflex*, 189–96. To be sure, this dispute had taken place in the 1830s and 1840s, and had mostly been settled by 1870. However, this is another example of mechanist tendencies in Huxley's questions; cf. Barton, *X Club*, 311–12.
49 T.H. Huxley to Henry Cole, 7 June 1861, in Cole, "Diaries," 1861, National Art Library (Great Britain), Manuscript MSL/1934/4117–4159.
50 Donnelly, *Instruction of the Industrial Classes in Elementary Science*, 11.
51 I am grateful to Elihu Gerson for teaching me this important point.
52 Richmond, "The 1909 Darwin Celebration."
53 Wallace, *My Life*, 2:20.
54 DSA, *Twenty-First Report* (1874), 22.
55 Wallace, *My Life*, 2:407–9, 416.
56 For the realist reader: to *temporarily* adopt a relativist, or agnostic, stance is a long-standing methodological principle in science and technology studies. This is because explaining that something occurred or was accepted because it was true, right, or real means that one stops asking how people come to accept or disbelieve it when that truth, rightness, or reality was still being established. Historians of science adopt this method to avoid hindsight and incuriosity. For a more contemporary explanation of this method – applied to understanding how physicists came to accept the existence of gravity waves in 2015 after a 50 year dispute – see Collins, *Gravity's Kiss*, 286-7.
57 Ansted, *The World We Live In*, 154.
58 Wallace, *My Life*, 2:411.
59 DSA, *Directory* (1872), 161.
60 Secord, "Knowledge in Transit," 661.
61 Secord, *Victorian Sensation*.

4. Daguerreotypes of the Mind

1 Warburton Committee, *Report on Medical Education* (1834), G. Guthrie (Q4924, 4962), Part 2:41-2, 53.
2 Warwick, "Mathematical World on Paper," 304–5; Warwick, *Masters of Theory*, 139–41.
3 This is Barry Barnes's definition, which was taken up by Theodore Porter in thinking about authority as power *minus* discretion. Barnes, *Nature of Power*, 58; Porter, "Quantification and the Accounting Ideal in Science," 642.

4 For one case see Philo-Veritas [pseud.], "Examination of a Student of the London University," 646.
5 Caswell, *The Art of Pluck*, 27–8.
6 Todhunter, *Essays on Subjects Connected with Education*, 42–3.
7 "Medical Studies!," 137–9.
8 Roscoe, *Sir Henry Enfield Roscoe*, 30.
9 Richard Jones to William Whewell, 16 November 1845; Trinity College Library, Add MS c52 105, qtd in Stray, "Shift from Oral to Written Examination," 43.
10 A First Class Man, "Viva Voce Examinations," *The Times*, 17 August 1855, 10.
11 University of London, *Minutes of the Senate* (1837–43), 30.
12 Daston and Galison, *Objectivity*.
13 [Mann], "Boston Grammar and Writing Schools," 334.
14 Reese, *Testing Wars*.
15 Goody, *Domestication of the Savage Mind*, 37, 78; Latour, "Visualization and Cognition," 20–2.
16 Warburton Committee, *Report on Medical Education*, G. Guthrie (Q4962), Part 2:53.
17 Gerson, "The Limits of Rationalized Coordination," 6–7; Gerson, "The American System of Research," 2:17–20.
18 Ezrahi, *The Descent of Icarus*, 36.
19 Quick, "Examination Papers," 97–100.
20 Hutchins, *Cognition in the Wild*.
21 University of London, *Minutes of the Senate* (1843–49), 2:26–7.
22 Craik, *Mr Hopkins' Men*, 92; Ball, *History of the Study of Mathematics at Cambridge*, 212–13; Clark, *Origins of the Research University*, 113–17, 119, 131–4; Goodwin, "The Sacrifice of Education," 318–20.
23 Goodwin, *Memoir of Bishop Mackenzie*, 52–4.
24 Babbage, *Passages from the Life of a Philosopher*, 436–7.
25 A Fresh Surgeon [pseud.], "New Examinations at the London College of Surgeons," 495; Select Committee on Medical Registration, *Report*, W. Lawrence (Q1744), 150–1; B. Brodie (Q1966–75), 67–9; Dale, *Present State of the Medical Profession*, 56–7.
26 Babbage, *On the Economy of Machinery and Manufactures*, 156–7.
27 Babbage, *Passages from the Life of a Philosopher*, 436–7. And the less-knowledgeable workers didn't have to be paid as much as the more knowledgeable ones; Babbage considered this to be one of his most important economic insights.
28 Light, "When Computers Were Women."
29 Agar, *The Government Machine*, 47–8, 51–4.
30 Agar, *The Government Machine*, 53–4, 60–2.
31 Porter, "Quantification and the Accounting Ideal in Science."

32 Committee of the Privy Council on Education, *Report* (1867–8), 297.
33 Committee of the Privy Council on Education, *Report* (1867–8), 296; on this recurring complaint about paper work not being "real" work see Lampland and Star, "Reckoning with Standards," 10–11.
34 Latham, *On the Action of Examinations*, 50.
35 Arnold, "The Twice-Revised Code," 360.

5. Machining Minds

1 See also Espeland and Stevens, "Commensuration," 314–16.
2 Ashworth, "Charles Babbage, John Herschel, and the Industrial Mind."
3 Daston and Galison, *Objectivity*.
4 *Re-Organisation of the Civil Service*, 114–15.
5 Edward, "Sir Charles Trevelyan and Civil Service Reform," 75.
6 Warwick, *Masters of Theory*, 174.
7 Taunton Commission, *Minutes of Evidence* I, W.B. Carpenter (Q747), 84–5.
8 Carteret-Bisson, *Our Schools and Colleges, for Boys*, 211.
9 *Re-Organisation of the Civil Service*, 62–3.
10 Espeland and Sauder, *Engines of Anxiety*, 29–30.
11 Espeland and Stevens, "Commensuration," 17–20, 314.
12 Kula, *Measures and Men*, 87.
13 Brock, "Geometry and the Universities."
14 Todhunter, *Elements of Euclid for the Use of Schools and Colleges*, vii; Brock, "Geometry and the Universities," 29.
15 T.B. Macaulay, 19 *Parliamentary Debates* H.C., 3rd ser. (1833), col. 526; see also Stray, *Classics Transformed*, 53.
16 Busch, *Standards*, 62–6; David, "Clio and the Economics of Qwerty."
17 Hacking, "Biopower and the Avalanche of Printed Numbers," 293.
18 Best, *Improvement of the Civil Service*, 13.
19 Booth, *How to Learn and What to Learn*, 20; Taunton Commission, *Minutes of Evidence* I, B. Price (Q628), 69–70.
20 There are at least three different spellings of this name in the official records.
21 University of London, *Minutes of the Senate* (1843–9), 2:3–6, 8, 9, 14–15, 62.
22 Trollope and Skilton, *An Autobiography*, 28–31.
23 Trollope, *The Three Clerks*, 2:247.
24 Qtd in Harte, *The University of London*, 1836–1986, 114.
25 University of London, *The Historical Record*, 26.
26 Tomlinson, "Admission of Females to the Examinations of the University," 3 July 1856, RC University of London Archive 19/1.
27 William B. Carpenter to Rt Hon Earl Granville, 8 May 1868, University of London Archive RC 19/7.

28 Taunton Commission, *Minutes of Evidence* I, W.B. Carpenter (Q959), 110. "Rut" is taken from Dr Storrar's question, and with "gymnastics," Carpenter is paraphrasing one of the medical senators.
29 Espeland and Stevens, "Commensuration," 327.
30 See, for instance, Romanes's argument in "Mental Differences between Men and Women."
31 Kuhn, *The Essential Tension*, 220; Schaffer, "Metrology, Metrication, and Victorian Values," 440–1; Agar, *Science in the Twentieth Century*.
32 Latham, *On the Action of Examinations*, 199–200; Melville, "The Prize System in Education," 249–50; Keetley, *The Student's Guide to the Medical Profession*, 23–4.
33 Booth, *Examination the Province of the State*, 21–2.
34 "Examinations," *Journal of the Society of Arts* 6, no. 286 (14 May 1858), 400–1.
35 "Letter to Secretary of Oldham Lyceum"; "Examiner's Report."
36 "Programme of Examinations for 1860."
37 C.U. Sohailels to Arthur Milman, 25 June 1880, University of London Archives RC 37/23.
38 William B. Carpenter to Sir Frederic Rogers, 17 September 1867, University of London Archives RO 1/2A/6; Carpenter to J.F. Elliot, 8 June 1868, University of London Archives RO 1/2A/7.
39 "Examinations Letter Book 1890–1891," Cambridge Assessment Archives A/LB 1/2.
40 William Garnett to Arthur Milman, 16 June 1884, University of London Archives RC 37/64.
41 University of London Examination Committee to William Garnett and A.W. Reinold, n.d., University of London Archives RO 2/13/9.
42 University of London, *Minutes of Committees* (1853–66), 152–4.
43 Magnus, qtd in Quick, "Examination Papers," 97–100.
44 Taunton Commission, *Minutes of Evidence* I, W.B. Carpenter (Q758–60), 86–7.
45 Taunton Commission, *Minutes of Evidence* I, P. Le Neve Foster (Q1212), 139–41.
46 Stray, "Shift from Oral to Written Examination," 44–5. On the relationship of series, numerical and otherwise, to individual symbols within that series, see Bourdieu, *Outline of a Theory of Practice*, 87–8.
47 "In a lump" was Price's phrase at the Taunton Commission, *Minutes of Evidence* I, B. Price (Q725), 79–81; see also University of Cambridge, *Examination of Students Who Are Not Members of the University*, 5.
48 Blackie, "Competitive Examinations for the Public Service," 319.
49 Boulger, *Candidate's Complete Instructor*, 55–7.
50 Alder, "Making Things the Same," 522–4; Alder, *The Measure of All Things*.

51 There must obviously be some caution as different categories are lumped together or different exams added. For instance, at Woolwich a comparison of the scheme between 1861 and 1884 shows what seems to have been a decline in the value of mathematics relative to Classics, but this conclusion is misleading – a preliminary exam in arithmetic was added, likely as an early "filter", yet no such equivalent exam in Classics existed. Hutchinson, *Guide to the Army-Competitive Examinations*, 15–17; *Fifteenth Report of Her Majesty's Civil Service Commissioners* (1870); Carteret-Bisson, *Our Schools and Colleges, for Boys*, 222.
52 Taunton Commission, *Minutes of Evidence* I, W.B. Carpenter (Q747), 82–4. Carpenter's assertion is confirmed in Committee of Examiners for Degrees in Arts (and Science), *Minutes*, 1853–67, University of London Archives RO 2/8/1, 2–3.
53 Taunton Commission, *Minutes of Evidence* I, W.B. Carpenter (Q759, 853), 86–7, 96.
54 Fitch, "Lectures on Practical Teaching," 307–8.
55 On definitions of a "gentleman," see Endersby, "Odd Man Out."
56 Taunton Commission, *Minutes of Evidence* I, B. Price (Q628), 70–1; (Liveing, Q225), 26.
57 Browne, *Recollections of a Bishop*, 72.
58 Cambridge Local Examination Syndicate, Minutes, 1862–7, Cambridge University Library, LES 1/1. Asterisks in original.
59 Taunton Commission, *Minutes of Evidence* I, B. Price (Q630), 70–1.
60 G. Bentinck, 158 *Parliamentary Debates* H.C., 3rd ser. (1860), col. 907.
61 Richard Harrison to University of London Registrar, December 1884, University of London Archives RC 8/5; Arthur Milman to F. Harrison, 30 January 1885, University of London Archives RO1/2A/19; F.V. Dickins to Richard Harrison, 23 December 1884, University of London Archives RO1/2A/19.
62 That is, in one year the pass rate was 55 per cent, and in another, 38 per cent, yet with few rule or topic changes that would explain the difference.
63 R.F. Weymouth et al., "Letter of Concern over Unevenly Marked Examinations," 5 February 1882, University of London Archives.
64 Edgeworth, "The Uncertainty of Examinations," 95.
65 Roach, *Public Examinations*, 283.
66 E. Arnold, "Church of England Schools Inspected in the Counties of Cornwall and Devon," 42–4.
67 Hopkinson, "The Arnolds and Elementary Education." Four of Thomas Arnold's five sons became school inspectors: in addition to Matthew and Edward, Thomas inspected schools in Van Diemen's land (Tasmania) and William in the Punjab.
68 Schaffer, "Modernity and Metrology," 71–2.

6. Thin Descriptions

1 Anne Wynne to John Tyndall, 9 July 1856, Royal Institution MS JT/1/TYP/5/1855–1856; Wynne to Tyndall, 18 March 1854, Royal Institution MS JT/1/TYP/5/1849.
2 Brock, *Queenwood College Revisited*, 15; John Tyndall to Thomas Archer Hirst, 21 February 1855, Royal Institution MS JT/1/T/HTYP/439.
3 Latour, *Science in Action*.
4 Schwartz, "Professions, Elites and Universities in England," 944; Collins, *The Credential Society*, 54–7; Weber, *Economy and Society*, 1:1000.
5 "History of the University of London," accessed 23 July 2020, https://london.ac.uk/about-us/history-university-london.
6 Harte, *University of London*, 12–13.
7 Harte, *University of London*, 64.
8 University of London, *Papers Relating to the Origin of the University*, 2.
9 Harte, *University of London*, 82–4.
10 Harte, *University of London*, 22–5, 36, 68–74, 138.
11 University of London, *The Historical Record*, 26.
12 Even this was too much for some: Matthew Arnold's father, Thomas, resigned from London's senate because he did not believe its degrees should be granted to Jews. Harte, *University of London*, 91.
13 Harte, *University of London*, 114.
14 University of London, *Minutes of the Senate* (1843–9), 2:62.
15 University of London, *Minutes of the Senate* (1859–62), 5:31, 33.
16 [Fitch], "London University Calendar for 1857," 20.
17 Post Office, *Jubilee of Uniform Inland Penny Postage*, 71–2; A.J. Mill, "Some Notes on Mill's Early Friendship with Henry Cole."
18 University of London, *The Historical Record*, ix.
19 Medical students had to present additional certificates.
20 R.W. Rothman to A. Mitchell, 10 June and 15 June 1839; Rothman to Mr. Baxter, 28 October 1839, University of London Archives RO 1/2/1.
21 R.W. Rothman to James Booth, 2 February 1841, University of London Archives RO 1/2/2.
22 R.W. Rothman to James Booth, 20 May 1842, University of London Archives RO 1/2/4. For Booth's time at Bristol see also Foden, *James Booth*, 29–33.
23 Harte, *University of London*, 96.
24 University of London, *Minutes of the Senate* (1843–9), 2:28.
25 University of London, *Minutes of the Senate* (1855–8), 4:80–8; the phrase "notoriously incompetent" is from [Fitch], "London University Calendar for 1857," 9, 12. Confusingly, there are two published versions of the senate minutes in the Senate House archives: this one (published by Taylor & Francis) and one for the years 1853 to 1858 published by Richard Taylor.

This overlap may have occurred because Richard Taylor joined William Francis in 1852 to form Taylor & Francis.

26 It is probably not unrelated that Rothman had by 1857 retired and been replaced by W.B. Carpenter as university registrar.

27 University of London, *Minutes of the Senate* (1853–8), 4:133.

28 University of London, *Minutes of the Senate* (1853–8), 4:90–2, 100, 137. See also [Fitch], "London University Calendar for 1857," 15–16.

29 Thomas Archer Hirst to John Tyndall, 20 November 1853, Royal Institution MS JT/1/H/185; Heinrich Debus to John Tyndall, 6 November 1853, Royal Institution MS JT/1/D/13; Hirst to Tyndall, 29 January 1854, Royal Institution MS JT/1/H/190. Under great pressure, Mummery resigned his Queenwood post.

30 University of London, *The Historical Record*, 46.

31 London's MB and MD degrees still required college certificates of attendance, recognizing practical instruction; nonetheless, there were still differences in course rigour between different affiliated colleges, meaning that the issue of "diversity of practice" would still have presented a problem.

32 Busch, *Standards*, 117.

33 University of London, *Minutes of the Senate* (1855–8), 4:52–65.

34 Froude, "Teaching English History," 51–2.

35 University of London, "Amended Draft Charter as Adopted by the Senate," June 4, 1857, University of London Archives RO 1/2/19, p. 13. It is known as the Charter of 1858 because this is the year when it was signed by the Queen and became official.

36 Harte, *University of London*, 106.

37 "Liverpool Petition for Medical Reform," 754; Dale, *Present State of the Medical Profession*, 2–3, 25–6.

38 "Scholastic Registration," 219. This did not succeed.

39 John Tyndall to John Tyndall Senior, 26 September 1840, Royal Institution MS JT/1/TYP/1604–1609; Tyndall, "Testimonials of John Tyndall."

40 Sylvester, *Lowe and Education*, 4.

41 Administrative Reform Association, *Appointments for Merit*, 9–10.

42 Chadwick, *Competitive Examinations for Admission to the Public Service*, 26. He also pointed out that testimonial letters were at times put to more dubious uses: a testimonial might be provided to "the greatest scamp of the parish in the hope of getting rid of him." In *Papers on the Re-organisation of the Civil Service*, 184.

43 "Registry of Candidates Who Have Gained Certificates," *Journal of the Society of Arts* 4, no. 200 (19 September 1856), 705. Booth was the journal's founder.

44 Booth, *Examination the Province of the State*, 18–20.

45 "Registry of Candidates Who Have Gained Certificates," 705.

46 "Examination of Members of Classes in Institutions," *Journal of the Society of Arts* 4, no. 190 (11 July 1856), 587.
47 Babbage, *On the Economy of Machinery and Manufactures*, 102–10.
48 *Re-Organisation of the Civil Service*, 193–4.
49 *Occasional Papers on University Matters*, 1–4.
50 Qtd in Harte, *University of London*, 116.
51 Rolleston and Turner, *Scientific Papers and Addresses*, l–li.
52 "Liverpool Petition for Medical Reform," 754.
53 Northcote, "Competitive Examination for the Civil Service," 279–80.
54 Booth, *How to Learn and What to Learn*; "First Ordinary Meeting," *Journal of the Society of Arts* 4, no. 209 (21 November 1856), 1, 7.
55 See for instance Fry, *Our Schools and Colleges*, 2nd edition, xvii–xviii; Barry, "The Good and Evil of Examination," 649–50.
56 Acland, *New Oxford Examinations*, 8.
57 "College of Preceptors – August Distribution of Prizes and Certificates," 215–16.
58 Taunton Commission, *Minutes of Evidence* I, John Robson (Q368), 42–3.
59 "College of Preceptors – August Distribution of Prizes and Certificates," 215–16.
60 Booth, *Examination the Province of the State*, 10–11; Acland, *New Oxford Examinations*, 8; "Educational Quacks"; J.S. Mill, "Endowments."
61 Desrosières, *History of Statistical Reasoning*, 13.
62 "Headmistresses' Reports to Governors 1871–1885," North London Collegiate School Archives RS 2ii.
63 Lowe, "Speech on the Revised Code," 240–1.
64 Mundella, "Proposals for a New Code," 332–3.
65 Porter, "Thin Description," 225.
66 Scott, *Seeing Like a State*, 77–8.
67 Agar, *The Government Machine*, 74–5.
68 Wiese, *German Letters on English Education*, 132.
69 Geertz, "Thick Description," 5, 13–14.
70 Porter, "Thin Description," 221–2.
71 To repeat, I am grateful to Simon Schaffer for this phrase.
72 Strathern, "Accountability ... and Ethnography."
73 Hacking, *Historical Ontology*, 108.
74 Playfair to Tyndall, 7 January 1857, Royal Institution MS JT/TYP/3/982–983.
75 University of London, *Minutes of the Senate* (1855–8), 4:52–65.
76 Booth, *Examination the Province of the State*, 26–8; Lowe, *Middle Class Education: Endowment or Free Trade*, 12.
77 "Reviews and Notices," *Educational Times*, May 1861, 40.
78 Fry, *Our Schools and Colleges*, 2nd edition, xvii–xviii. Fry's greater reliance on Local Examination results between the first (1867) and this second (1868) edition of his *Directory* is notable.

79 Taunton Commission, *Minutes of Evidence* I, G.D. Liveing (Q141), 16–17.
80 Taylor, "Cambridge Local Examinations."
81 Carteret-Bisson, *Oxford and Cambridge Local Examination Record*, 97.
82 Fry, *Our Schools and Colleges*, xvii.
83 Espeland and Sauder, "Rankings and Reactivity," 21.

7. Learning and Earning

1 Cheth [pseud.], "A Competitive Examination," 108–9.
2 Cheth, "A Competitive Examination," 71–4.
3 Latham, *On the Action of Examinations*, 9.
4 Stephen, *Sketches from Cambridge*, 55.
5 Russell, *Portraits from Memory*, 15–16.
6 Wilson, *Autobiography*, 48–9.
7 At the time, only coaches were known as "crammers"; one didn't "cram" oneself.
8 "Experiences of a 'Little-Go' Coach," 6–7.
9 Fitch, "Lectures on Practical Teaching," 307–8.
10 "Oxford Local Examination," January 1864, 221.
11 Faulkner and Becker, *Do You Know: The Jazz Repertoire in Action*, 190–1.
12 Reathlous, "Matriculation Hints," 567–8; Workman, "How to Write an Examination Paper."
13 Examinee [pseud.], "Notes in the Senate House," 81–3.
14 The Chinese Room is a puzzle akin to the Turing Test to think about artificial intelligence. It was created by the philosopher John Searle: if a person ignorant of Chinese sits in a sealed room and has only a book and extremely detailed grammatical rules of its use, could this person feign knowledge of Chinese, responding to someone outside the room sending messages in Chinese?
15 Thomas Archer Hirst to John Tyndall, 16 November 1856, Royal Institution MS JT/1/HTYP/480–481. Italics in original.
16 John Tyndall, "Journals," 1855–72, Journal VIIIa, November 1859, 1165.
17 Todhunter, *Essays on Subjects Connected with Education*, 37.
18 Taunton Commission, *Minutes of Evidence* II, 650.
19 Eddy, "The Interactive Notebook," 91–2; Eddy, "The Shape of Knowledge," 223–4, 228–9.
20 Epictetus, *Discourses and Selected Writings*, 242.
21 Locke, "Of the Conduct of the Understanding," 356; Dodds, *Matriculation at the University of London*, 8–9; Playfair, *On Teaching Universities*, 26–7.
22 Lakoff and Johnson, *Metaphors We Live By*, 25–9, 53–5, 147–9.
23 "A Word for 'Cram,'" *London Review*, 18 August 1866, 179.
24 DSA, *Eighteenth Report* (1871), 51–2; Stephen, *Sketches from Cambridge*, 55.
25 Trollope, *The Three Clerks*, 1:236.

26 Sayce, "Results of the Examination-System," 838; Wilson, *Autobiography*, 48–9, 64; Warwick, *Masters of Theory*, 188.
27 Playfair, *Subjects of Social Welfare*, 286–7.
28 *Second Report of the Royal Commissioners on Technical Instruction*, T.H. Huxley (Q3007), 3:325.
29 T.H. Huxley and L. Huxley, *Life and Letters of Thomas Henry Huxley*, 3:462.
30 Desmond, *Huxley*, 14–15.
31 T.H. Huxley, "Technical Education [1877]," 422.
32 Cole and Cole, *Fifty Years of Public Work*, 1:311–12; Samuelson Committee, *Report*, 401.
33 Cheth, "A Competitive Examination," 107.
34 Dale, *Present State of the Medical Profession*, 57–8; Stray, *Grinders and Grammars*.
35 Taunton Commission, *Minutes of Evidence* II, 650.
36 Herbert George Wells to *Saturday Review* editor, 10 December 1895, qtd in Smith, *Correspondence of H.G. Wells*, 1:251.
37 Warwick, *Masters of Theory*, 89–91.
38 Appleyard, *Letters from Cambridge*, 67–9.
39 Foden, *Philip Magnus: Victorian Educational Pioneer*, 45–7.
40 Clarke, *The Influence of Pass Examinations*, 11–12. For specific details of Oxford coaching, see Oxonian [pseud.], "Almae Matres," 577–9.
41 Stephen, *Sketches from Cambridge*, 45, 102.
42 Stephen, *Sketches from Cambridge*, 100–1.
43 "Mr. Golightly," 339–41. The curious reader may consult Paley, Works (pp. 69–75) to see how accurately this mnemonic helped students recall this section.
44 Taunton Commission, *Minutes of Evidence* I, R. Moseley (Q1906), 195.
45 Select Committee on Medical Registration, *Report*, 179–80.
46 "Medical Studies!," 137–9.
47 Roach, *Public Examinations*, 269.
48 "Obituary – Mr. Walter Wren," 4; Besant, "The Late Walter Wren," 8; Papillon, "The Late Walter Wren," 10; "Life at a 'Crammer's,'" 139–40; Indian Civil Service, *Selection and Training of Candidates*, 298.
49 Wren lost his seat due to a funding scandal not linked to his coaching activities (and that occurred without his knowledge).
50 "Dispute between 'Crammers' and Public Schools," 44, 61; Wolfram, The Private Tutor's "Raison D'être," 6–8.
51 Besant, "The Late Walter Wren," 8.
52 Indian Civil Service, *Selection and Training of Candidates*, 40.
53 Taunton Commission, *Minutes of Evidence* II, Mark Pattison (Q17872), 952.
54 Suum Cuique [pseud.], "The Indian Civil Service," *The Times*, 6 August 1862, 6.

55 Farnell, *The Examination*, 30–3, 94–6, 100–2.
56 "Educational Summary of the Month," *Educational Times*, October 1879, 282–3.
57 Contributions would be regarded as "STRICTLY PRIVATE" and the district in which the questions were set would not be disclosed, for "obvious reasons." "Recent Examination Questions,"47.
58 Fitch, "Statistical Fallacies Respecting Public Instruction," 620. This practice was ended in the 1870 revision of the Revised Code.
59 Farnell, *The Examination*, 104–6.
60 Frank Foden suggests that the DSA rules resembled the complicated Queen's regulations because Donnelly came from the Royal Engineers. Foden, *Philip Magnus: Victorian Educational Pioneer*, 145.
61 DSA, *Eighteenth Report* (1871), 79–82.
62 Sherborne, *H.G. Wells*, 38–40, 42–3.
63 H.G. Wells to Sarah Wells, 17 July 1883, *Correspondence*, 1:30.
64 Sherborne, *H.G. Wells*, 48–9; Wells, *Love and Mr. Lewisham*, 17–18.
65 Wells, *Love and Mr. Lewisham*, 68–9; Wells, *Autobiography*, 137–8; H.G. Wells to Sarah Wells, 17 July 1884, *Correspondence*, 1:34; Sherborne, *H.G. Wells*, 51–2.
66 Wells, *Love and Mr. Lewisham*, 68–9; Sherborne, *H.G. Wells*, 54–5.
67 One of Wells's tools was seen as so notoriously bad it was kept in a display case as a warning to others: Wells, *Autobiography*. But one of Wells's biographers notes that Wells left being a draper's apprentice to avoid manual labour; Sherborne, *H.G. Wells*, 58–9. For Wells, then, science was "headwork," an interesting contrast with the twentieth-century emphasis in science and technology studies of the dependence of science upon tacit knowledge.
68 Wells, Autobiography, 261–3; Wells, *The New Machiavelli*, 20, 25–7.
69 H.G. Wells to A.T. Simmons, 18 September 1889, *Correspondence*, 1:128; Wells to Simmons, Spring 1890, *Correspondence*, 1:144.
70 Sherborne, *H.G. Wells*, 73; Wells, *Autobiography*, 275–6.
71 Wells, *Autobiography*, 284–5.
72 Wells to Simmons, Spring 1890, *Correspondence*, 1:144.
73 Wells, *Autobiography*, 143, 275–6; Sherborne, *H.G. Wells*, 75; de Salvo, *University Correspondence College*, 13.
74 Heller, *London Clerical Workers*, 171.
75 "Well-Known Teachers at Work: William Briggs," 384.
76 "Prospectus of University Correspondence College," 2.
77 The phrase "reverse engineering" is used here as Briggs's work seems similar to what contemporary deans do in order to raise their law schools up the *U.S. News and World Report* rankings: Espeland and Sauder, *Engines of Anxiety*, 35.
78 Wells, *Autobiography*, 281–2; "Well-Known Teachers at Work: William Briggs," 386–7.

79 de Salvo, *University Correspondence College*, 13–14; Wells, *Text-Book of Biology*; Sherborne, *H.G. Wells*, 75.
80 "Well-Known Teachers at Work: William Briggs," 387.
81 de Salvo, *University Correspondence College*, 13; Sherborne, *H.G. Wells*, 76–7; Wells, *Text-Book of Biology*.
82 "Well-Known Teachers at Work: William Briggs," 387–8; Wells, *Text-Book of Biology*.
83 Wells, *Autobiography*, 284–5.
84 Wells, *Text-Book of Biology*, viii–ix; Wells, *Autobiography*, 35–6, 133–4.
85 de Salvo, *University Correspondence College*, 33.
86 de Salvo, *University Correspondence College*, 13; Briggs, *Intermediate Mathematics*, viii; "Prospectus of University Correspondence College," 4, 13; Wells, *Autobiography*, 281–2; "Well-Known Teachers at Work: William Briggs," 385–6.
87 Wells, *Text-Book of Biology*, 140–9; "Prospectus of University Correspondence College," 13.
88 Henry J. Thomson to University of London Registrar, 31 October 1893, University of London Archives RC 8/37; F.V. Dickins to Alexander Thomson, 10 November 1893, University of London Archives RO1/2A/22.
89 "Advertisements," vi; Wells, *Autobiography*, 285.
90 Wells, *Autobiography*, 285.
91 Wells, *Autobiography*, 286, 290.
92 Yuan, "On the Funeral of Mr Sun Yat-sen," 4–5. My thanks to Hsiang-Fu Huang for this reference.

8. Immoral Economies

1 Select Committee, *Mr. Goffin's Certificate*, J.F.D. Donnelly (Q7–10, 33–85, 196), 1–2, 6–7, 24; R.E.H. Goffin (Q1990), 126; R.R. Barnett (Q480–514), 203; Appendix 1, 180.
2 Poovey, *Genres of the Credit Economy*, 5–6.
3 Espeland and Sauder, *Engines of Anxiety*, 142.
4 Porter, "Thin Description," 225.
5 *The Wire*, season 4, episode 9, "Know Your Place," directed by Alex Zakrzewski, written by David Simon, aired 12 November 2006 on Home Box Office.
6 Select Committee, *Mr. Goffin's Certificate*, Appendix 6, 263.
7 Samuelson, "The Progress of Science Schools," 228.
8 Samuelson, "The Progress of Science Schools," 227; "Rewards and Honours for Proficiency in Science," 126; Butterworth, "Science and Art Department," 369–70.

9 Select Committee, *Mr. Goffin's Certificate*, Appendix 1, 190.
10 Samuelson, "The Progress of Science Schools," 226, 28, 28n.
11 Select Committee, *Mr. Goffin's Certificate*, Appendix 3, 214–15; G. Hamilton, 264 *Parliamentary Debates* H.C., 3rd ser. (1881), cols. 1270–1.
12 Select Committee, *Mr. Goffin's Certificate*, Appendix 6, 261–3.
13 DSA, *Eighteenth Report* (1871), 129.
14 Select Committee, *Mr. Goffin's Certificate*, Appendix 6, 264–5; DSA, *Eighteenth Report* (1871), 29, 69, 86, 100.
15 Select Committee, *Mr. Goffin's Certificate*, C. Ledger (Q1487–8), 90.
16 Select Committee, *Mr. Goffin's Certificate*, C. Ledger (Q1555, 1559), 93; H. Brummell (Q1776–80, 1785–9), 102–3, 1491; Appendix 5, 233–41.
17 Select Committee, *Mr. Goffin's Certificate*, Appendix 3, 216; Butterworth, "Science and Art Department," 196, 369–71; G. Hamilton, 264 *Parliamentary Debates* H.C., 3rd ser. (1881), col. 1270; DSA, *Twentieth Report* (1873), 3.
18 *Re-Organisation of the Civil Service*, 5–6.
19 Bede [and Bradley], *The Adventures of Mr. Verdant Green*, 98–9, 103.
20 Thorold, *The Life of Henry Labouchere*, 4, 23–7.
21 *The Times*, 15 January 1855, 4.
22 "College of Preceptors: Dean's Report," 245; Cambridge Local Examination Syndicate, Local Examinations (1866), 1–2; Cambridge Local Examination Syndicate, "Minutes" (1868–85), Cambridge University Library, LES 1/2; Quick, "Examination Papers," 97–100.
23 University of London, *Minutes of the Senate* (1879–82), vol. 10:20, 24, 90, 95, 98, 114–15; Vincent B. Lewes to Thomas Douse, 25 August 1880, University of London Archives RC 40/20; Thomas Savage to Arthur Milman, 27 August 1880, University of London Archives RC 40/20; J.M. Hill to Arthur Milman, 30 August 1880, University of London Archives RC 40/20; Heinrich Debus to Arthur Milman, 18 October 1880, University of London Archives RC 40/20.
24 William B. Carpenter to C. Brace, 11 July 1873, University of London Archives RO1/2A/10.
25 University of London, *Minutes of the Senate* (1843–9), 2:61–2 (19 July 1848).
26 University of London, Minutes of Committees (1853–66), University of London Archives ST 3/2/6, 10–11.
27 *The Times*, 31 May 1872, 12.
28 The report does not say which location.
29 DSA, *Fifteenth Report* (1868), 1–2, 79; Butterworth, "Science and Art Department," 200; DSA, *Twenty-Second Report* (1875), 380–3; DSA, *Eighteenth Report* (1871), 30, 79; DSA, *Ninteenth Report* (1872), 7, 9; Butterworth, "Science and Art Department Examinations," 38–9.

30 Butterworth, "Science and Art Department," 369–70; S. Waterlow, G. Hamilton, 264 *Parliamentary Debates* H.C., 3rd ser. (1881), col. 1287; Select Committee, *Mr. Goffin's Certificate*, Appendix 6, 264–5.
31 United Westminster Schools *Annual* (1880), 6.
32 "Science Training in Elementary Schools," 63.
33 Select Committee, *Mr. Goffin's Certificate*, C. Ledger (Q1595–8), 95.
34 Select Committee, *Mr. Goffin's Certificate*, C. Ledger (Q1567–72, 1596–7), 93–5; Butterworth, "Science and Art Department," 369–70.
35 Select Committee, *Mr. Goffin's Certificate*, J.F.D. Donnelly (Q147–8), 19. Although Donnelly was unsure if payments went straight to Goffin or to the governing board, we do know it went to the board since if Goffin had been paid directly by the DSA, they would have been able to charge Goffin with fraud.
36 Select Committee, *Mr. Goffin's Certificate*, J.M. Coates (Q480–99), 49–50.
37 Select Committee, *Mr. Goffin's Certificate*, C. Ledger (Q1593, 1604–18), 95, 96.
38 Select Committee, *Mr. Goffin's Certificate*, J.F.D. Donnelly (Q138–9), 18; DSA, "General Order Book," 1857–1904, National Art Museum 86-AA-30.
39 Stray, "Educational Publishing," 505.
40 Select Committee, *Mr. Goffin's Certificate*, J.F.D. Donnelly (Q126–32, 138, 140–6), 17–18; J.F.D. Donnelly and W. Abney (Q373–7), 40.
41 Select Committee, *Mr. Goffin's Certificate*, J.F.D. Donnelly (Q117), 16.
42 "The Goffin Case," *Educational Times*, 1 October 1879, 284.
43 Select Committee, *Mr. Goffin's Certificate*, J.F.D. Donnelly and W. Abney (Q377–85), 40–1.
44 Select Committee, *Mr. Goffin's Certificate*, R. Jones (Q2343–69, 2519–35), 148–9, 156; J.F.D. Donnelly (Q22–46, 153–9), 3–4, 19–20; F.T. Davies (1212–1318), 78–82; Appendix 1, 174–6.
45 Select Committee, *Mr. Goffin's Certificate*, Appendix 6, 243.
46 G. Hamilton, 264 *Parliamentary Debates* H.C., 3rd ser. (1881), col. 1274; Select Committee, *Mr. Goffin's Certificate*, J.F.D. Donnelly (Q51, 59, 74, 88, 91–5, 163), 4–6, 9–10, 20; J.F.D. Donnelly and W. Abney (Q236–42, 346–9), 28–9, 37.
47 Select Committee, *Mr. Goffin's Certificate*, Appendix 7, 280.
48 Select Committee, *Mr. Goffin's Certificate*, Appendix 7, 280.
49 Select Committee, *Mr. Goffin's Certificate*, J.M. Coates (Q552–4, 612–33), 52, 55–6.
50 Select Committee, *Mr. Goffin's Certificate*, Appendix 2, "Minutes of an Inquiry at the United Westminster Schools," 192–213.
51 Select Committee, *Mr. Goffin's Certificate*, J.F.D. Donnelly (Q195–7), 23–4; Appendix 1, 183–7; United Westminster Schools, *Annual*, 32–5.
52 Select Committee, *Mr. Goffin's Certificate*, J.M. Coates (Q555, 638), 53, 56.

53 G. Hamilton, 264 *Parliamentary Debates* H.C., 3rd ser. (1881), col. 1271.
54 Select Committee, *Mr. Goffin's Certificate*, C. Ledger (Q1504–39), 90–2; R.E.H. Goffin (Q2071–108), 134–6; R. Jones (Q2344–7), 148; G. Hamilton, 264 *Parliamentary Debates* H.C., 3rd ser. (1881), cols. 1273–4.
55 Select Committee, *Mr. Goffin's Certificate*, iii.
56 Butterworth, "Science and Art Department," 372.
57 The case was precedent setting because Donnelly's testimony was before a parliamentary committee, which meant it was covered by parliamentary privilege.
58 G. Hamilton, 264 *Parliamentary Debates* H.C., 3rd ser. (1881), cols. 1276–7; S. Waterlow, col. 1285; Butterworth, "Science and Art Department," 372–3.
59 Select Committee, *Mr. Goffin's Certificate*, R.E.H. Goffin (Q2005–9), 128; Appendix 6, 255–8.
60 F. O'Donnell, 264 *Parliamentary Debates* H.C., 3rd ser. (1881), cols. 1296–8; S. Waterlow, col. 1285.
61 Butterworth, "Science and Art Department," 201; DSA, *Thirty-First Report* (1881), 61.
62 "Educational Summary of the Month," November 1880, 279.
63 G. Hamilton, 264 *Parliamentary Debates* H.C., 3rd ser. (1881), cols. 1277–8.
64 S. Waterlow, 264 *Parliamentary Debates* H.C., 3rd ser. (1881), col. 1284.
65 "Science Training in Elementary Schools," 63.
66 C. Warton, 264 *Parliamentary Debates* H.C., 3rd ser. (1881), cols. 1280–1.
67 "Examinations Letter Book 1892–1894," Cambridge Assessment Archives A/LB 1/3.
68 Rachel Aviv, "Wrong Answer," *The New Yorker*, 21 July 2014, accessed 26 July 2020, https://www.newyorker.com/magazine/2014/07/21/wrong-answer.

9. Economies, Remoralized

1 Stray, "Zulu and Zuleika." "Spooni" is likely a play on the name given to the lowest-placing honours candidate on the Mathematics Tripos, who became known as the "wooden spoon."
2 Seivad [Davies], *The Moslem in Cambridge*, 1–2. See also Jackson, "Zuleika and the Moslem"; Stray, "Zulu and Zuleika."
3 Seivad [Davies], *The Moslem in Cambridge*, 1–2.
4 Espeland and Stevens, "Commensuration," 330.
5 Emily Davies to Henry Jackson, 27 July 1887, Girton College Archives, qtd in McWilliams-Tullberg, *Women at Cambridge*, 95.
6 [Burrows], "Schools Inquiry Commission," 456.
7 [Bryant], "Examinations as an Aid to School Work," 104.

8 Jex-Blake, "The Medical Education of Women," 391–2; see also her *Medical Women*.
9 For instance, see [R. Wilson], "Aesculapia Victrix," 23; Clough, "The Progress of Woman"; Scharlieb, "The Progress of Woman in Science and Medicine"; [Beale], "Girls' Schools"; An Oxford BA [Chisholm], "University Degrees for Women"; The petition is detailed in Herbert, *The Sacrifice of Education to Examinations*.
10 "What Women Are Educated For," 175.
11 Harris, *Common Threads*, 79–81.
12 Dickens, *Sketches by Boz*, 322; Taunton Commission, *Report*, 1:548–9.
13 Pedersen, *Reform of Girls' Education*, iii, 46–7.
14 Taunton Commission, *Minutes of Evidence* II, E.E. Smith (Q15718), 698; A Lady Principal, "Life and Work in a Girls' School," 247–8; Bodichon, "Middle-Class Schools for Girls," 432–3; Beale, *On the Education of Girls*, 3. See also Shteir, *Cultivating Women, Cultivating Science*, 228–9.
15 Pedersen, *Reform of Girls' Secondarry and Higher Education*, 46–7, 52–3.
16 Taunton Commission, *Minutes of Evidence* II, E. Davies (Q11214), 232.
17 Wiese, *German Letters on English Education*, 132. See also Pedersen, *Reform of Girls' Secondary and Higher Education*, 52–3.
18 Even these exams were not seen as prestigious enough, however: the pioneering headmistress of Cheltenham School for Girls, Dorothea Beale, testified that the Local exams of Oxford and Cambridge were looked down upon at her school since they had been established for boys of lower social status who did not intend to go to university. The girls who attended Cheltenham, however, had brothers who did intend to go to university.
19 "Proposed Admission of Girls to University Local Examinations," January 1864, North London Collegiate School Archives, Album 2: 1850–1876.
20 Pascoe, *Schools for Girls and Colleges for Women*, 33–4.
21 Frances Martin to Alexander Macmillan, 11 May 1868, Macmillan Archives, British Library Add. MS 54974 f39.
22 Taunton Commission, *Minutes of Evidence* II, Frances Martin (Q15494), 683.
23 Espeland and Sauder, "Rankings and Reactivity."
24 Not necessarily because of bias, but because of the senate convention that tying votes broken by the vice chancellor ought to preserve the existing order.
25 Harte, *University of London*, 114–16; [Eastlake], "Commission on the Education of Girls," 58; Howarth, "Introduction," xxv.
26 One person, possibly Dorothea Beale, thought that many questions on London's special certificate exam were more difficult than the Matriculation's questions in the same subject. Becker, "On the Study of Science by Women," 401–2.

27 W.G. Adams and G.C. Foster, "Application for Increase in Salary and for Appointment of Assistant Examiners," 13 January 1871, University of London Archives RC 37/7.

28 Barton, *The X Club*, 345.

29 See the similar wording of Queen's Institute of Female Professional Schools, "Respectful Memorial of the Committee of the Queen's Institute," 22 April 1874; Rugby Council for Promoting of Education for Women; Leeds Association of Schoolmistresses, May 1874; and the Central Committee of the National Union for Improving the Education of Women, all in University of London Archive RC 19/10a, b, c, and d. Three of the petition's signatures were from London's own senate members.

30 [Eastlake], "Commission on the Education of Girls," 58.

31 London Graduates in Medicine, "Memorial against Admitting Women to Degrees," 18 April 1877, University of London Archives RC 19/15; Wilks, "Admission of Women to the University of London"; see also Harte, *University of London*, 26–7, 114–16.

32 Tilly, *Social Movements*, 4.

33 Pedersen, *Reform of Girls' Secondary and Higher Education*, 99–100.

34 Sutherland, "Anne Jemima Clough and Blanche Athena Clough," 106–7; Sutherland, *Faith, Duty and the Power of Mind*, 88.

35 Davies, *Special Systems of Education for Women*, 3, 11.

36 Cambridge Local Examination Syndicate, "Minutes" (1862–7), Cambridge University Library, LES 1/1. Despite the historical significance of this decision, there is surprisingly little discussion to be found in Syndicate minutes on this matter.

37 "Proposed Admission of Girls to University Local Examinations (1)," January 1864, North London Collegiate School Archives, Album 2: 1850–1876.

38 Emily Davies to Frances M. Buss, 8 March 1864, North London Collegiate School Archives, Album 2: 1850–1876; Taunton Commission, *Minutes of Evidence* II, E. Davies (Q11310, 11311), 241–2.

39 Cambridge Local Examination Syndicate, *Report on an Examination of Girls*, 2, 11; Sutherland, *Faith, Duty and the Power of Mind*, 82; Davies, *Special Systems of Education for Women*, 10.

40 "Report of a Discussion on the Proposed Admission of Girls to the University Local Examinations," 2.

41 See, for instance, Romanes, "Mental Differences between Men and Women," 655–7; Thorburn, *Female Education from a Physiological Point of View; The Nonconformist & Independent*, 10 November 1881, clipping in North London Collegiate School for Girls Archives Album 1, 1877–1914.

42 Around mid-century a general and often-implicit belief was that the physiology of the reproductive system was opposed to the physiology of the nervous system: the fixed amount of energy held by each person

could be directed either toward sex or the mind. In such an equation, a woman focusing too much on cultivating her mind might thereby lose her femininity. Farley, *Gametes and Spores*.

43 The exam table can be found in in Album 2: 1850–1876 at the North London Collegiate School Archives; Cumberlege, *The North London Collegiate School 1850–1950*, 73–4.

44 Stray and Sutherland, "Mass Markets: Education," 371.

45 Taunton Commission, *Minutes of Evidence* II, Emily Davies (Q11300–1), 240; Davies, *The Application of Funds to the Education of Girls*, 3–4.

46 Howarth, "Introduction"; Sutherland, "Emily Davies, the Sidgwicks and the Education of Women in Cambridge," 40.

47 Stoddart, *Life and Letters of Hannah E. Pipe*, 179.

48 Eschbach, *The Higher Education of Women*, 72.

49 Sutherland, "Anne Jemima Clough and Blanche Athena Clough," 106–7.

50 Besant, "The Mathematical Tripos."

51 Jones, "Femininity and Mathematics at Cambridge circa 1900," 150.

52 Gardner, *A Short History of Newnham College*, 42.

53 "Girton College," 161.

54 *The Times*, 23 April 1880.

55 Roach, *Public Examinations*, 215–16.

56 Sutherland, *Faith, Duty and the Power of Mind*, 106–7.

57 Sutherland, *Faith, Duty and the Power of Mind*, 125; Siklos, *Philippa Fawcett and the Mathematical Tripos*, 27–8.

58 Cumberlege, *The North London Collegiate School 1850–1950*, 62–4.

59 Jones, "Femininity and Mathematics at Cambridge Circa 1900," 58, 61–4, 156; McWilliams-Tullberg, *Women at Cambridge*, 102–3.

60 This is noted in the case of Mary Adamson's experiences at UCL in Dyhouse, *No Distinction of Sex? Women in British Universities 1870–1939*, 33–4. In "A Lab of One's Own," Marsha L. Richmond studies the "female subcultures" in science at Cambridge that lingered until 1948.

61 Again, I thank Marsha Richmond for her this useful metaphor about "pushing" and "pulling."

62 Sutherland, "Emily Davies, the Sidgwicks and the Education of Women in Cambridge," 45.

63 Friedman, *The University of Toronto*, 87.

64 Porter, "Quantification and the Accounting Ideal in Science," 639–42.

65 Hoskin, "The Examination"; Hoskin and Macve, "Accounting and the Examination"; Foucault, *Discipline and Punish*, 189; Porter, "Quantification and the Accounting Ideal in Science," 642.

66 James Matthews Duncan to Arthur Milman, November 1882, University of London Archives RC 37.

Conclusion

1 Foucault, *Society Must Be Defended*, 29.
2 Latour, "Visualization and Cognition"; Latour, *Pandora's Hope*, chapter 2, "Circulating Reference: Sampling the Soil in the Amazon Forest."
3 Mirowski, "Looking for Those Natural Numbers," 166–7. The same argument is made in by Adrian Johns in *The Nature of the Book*.
4 Wright Mills, *The Power Elite*, via Lawrence Busch, *Standards: Recipes for Reality*, 30.
5 Pollock, *Oxford Lectures, and Other Discourses*, 219–20.
6 Taunton Commission, *Minutes of Evidence* I, R. Moseley (Q1862–8), 191–2.
7 Galton, *Hereditary Genius*, 19–21.
8 Renwick, "The Practice of Spencerian Science," 352; Rose, "Calculable Minds," 193–4.
9 Gigerenzer, "From Tools to Theories," 255.
10 "Brain-Power," 650–1.
11 Faulkner and Becker, *Do You Know: The Jazz Repertoire in Action*, 190–1.

Bibliography

Archival Sources

British Library
Gladstone Papers, Add. MS 44,333
MacMillan Archives, Add. MS 54794 f39
Cambridge Assessment
Archives
Cambridge University Library
Cambridge Local Examination Syndicate, LES
Imperial College London
College Archives
Huxley Papers
National Art Gallery (Great Britain)
Cole Collection
Cole Correspondence
Department of Science and Art
North London Collegiate School
Archives
Royal Institution, London
Correspondence of John Tyndall
Journal of John Tyndall
University of London, Senate House, London
Papers of the Senate, Minutes of Senate and Senate Committees
Registrar's Office

Published Sources

Acland, Thomas Dyke. *Some Account of the Origin and Objects of the New Oxford Examinations*. 2nd ed. London: J. Ridgway, 1858.

– *Middle Class Education: West of England Examination Papers, June, 1857*. Vol. 2. Exeter: W.H. Roberts, 1857.

Administrative Reform Association. *Appointments for Merit Discussed in Official Answers to Official Objections to the Abolition of Patronage*. London: M.S. Rickerby, 1855.

– *Unfitness of the Present Home Government for the Performance of New and Important Public Duties*. London: M.S. Rickerby, 1855.

"Advertisements." *The Practical Teacher* 1 (1881): i–xx.

Agar, Jon. *The Government Machine*. Cambridge, MA: MIT Press, 2003.

– *Science in the Twentieth Century and Beyond*. Cambridge: Polity, 2012.

Alborn, Timothy. "Quill-Driving: British Life-Insurance Clerks and Occupational Mobility, 1800–1914." *Business History Review* 82 (2008): 31–58.

[Alcock, Rutherford]. "The Peking Gazette." *Fraser's Magazine* 7 (1873): 341.

Alcock, Thomas. *Questions on Huxley's Lessons in Elementary Physiology*. London: Macmillan, 1868.

Alder, Ken. "Making Things the Same: Representation, Tolerance, and the End of the Ancien Regime in France." *Social Studies of Science* 28, no. 4 (1998): 499–545.

– *The Measure of All Things: The Seven-Year Odyssey and Hidden Error That Transformed the World*. New York: Free Press, 2002.

– "A Revolution to Measure: The Political Economy of the Metric System in France." In *The Values of Precision*, edited by Norton Wise, 39–71. Princeton: Princeton University Press, 1995.

Almond, Hely Hutchinson. *Mr. Lowe's Educational Theories Examined from a Practical Point of View*. Edinburgh: Edmonston & Douglas, 1868.

Ambirajan, Srinivasa. "Steam Intellect and the Raj: South India in the Nineteenth Century." In *The Steam Intellect Societies: Essays on Culture, Education and Industry circa 1820–1914*, edited by Ian Inkster, 160–80. Nottingham: Department of Adult Education, University of Nottingham, 1985.

"Analyses of Books, Experimental Chemistry for Junior Students." *Journal of Science* 6 (April 1884): 240.

Ansted, David Thomas. *The World We Live in; or, First Lessons in Physical Geography*. London: W.H. Allen, 1868.

Appleyard, Ernest Silvanus. *Letters from Cambridge, Illustrative of the Studies, Habits, and Peculiarities of the University*. London: John Richardson, 1828.

Armstrong, Henry Edward. "Chemistry." In *Chapters on the Aims and Practice of Teaching*, edited by Frederic Spencer, 222–59. Cambridge: Cambridge University Press, 1897.

Armytage, Walter Harry Green. "J.F.D. Donnelly: Pioneer in Vocational Education." *The Vocational Aspect of Education* 2, no. 4 (1950): 6–21.

Arnold, Edward. "General Report, for the Year 1867, on the Church of England Schools Inspected in the Counties of Cornwall and Devon." In *Report of the Committee of Council of Education, 1867–8*, 35–53. London: Eyre and Spottiswoode, 1868.

Arnold, Matthew. "The Twice-Revised Code." *Fraser's Magazine* 65 (1862): 347–65.

Arnold, Matthew, and George William Erskine Russell. *Letters of Matthew Arnold, 1848–1888*. London: Macmillan, 1895.

Ashworth, William. "Memory, Efficiency, and Symbolic Analysis: Charles Babbage, John Herschel, and the Industrial Mind." *Isis: International Review Devoted to the History of Science and Its Cultural Influences* 87 (1996): 629–53.

Babbage, Charles. *On the Economy of Machinery and Manufactures*. London: Charles Knight, 1832.

– *Passages from the Life of a Philosopher*. London: Longman, 1864.

Baldwin, Melinda. *Making Nature: the History of a Scientific Journal*. Chicago: University of Chicago Press, 2015.

Ball, Nancy. *Her Majesty's Inspectorate 1839–1849*. Edinburgh: Oliver and Boyd, 1963.

Ball, Walter William Rouse. *A History of the Study of Mathematics at Cambridge*. Cambridge: University Press, 1889.

Barnes, Barry. *The Nature of Power*. Cambridge: Polity, 1988.

Barry, Alfred. "The Good and Evil of Examination." *Nineteenth Century* 3, no. 14 (1878): 647–67.

Barton, Ruth. "'Huxley, Lubbock, and Half a Dozen Others': Professionals and Gentlemen in the Formation of the X Club, 1851–1864." *Isis* 89 (1998): 410–44.

– *The X Club: Power and Authority in Victorian Science*. Chicago: University of Chicago Press, 2018.

[Beale, Dorothea]. "Girls' Schools, Past and Present." *Nineteenth Century* 23 (1888): 541–54.

Beale, Dorothea. *On the Education of Girls*. London: Bell & Daldy, 1866.

Becker, Lydia Ernestine. "On the Study of Science by Women." *Contemporary Review* 10 (1869): 386–404.

Becker, Peter, and William Clark, eds. *Little Tools of Knowledge: Historical Essays on Academic and Bureaucratic Practices*. Ann Arbor: University of Michigan Press, 2001.

Bede, Cuthbert [and Edward Bradley]. *The Adventures of Mr. Verdant Green*. New York: Carleton, 1878.

Benson, Arthur Christopher, and Reginald Baliol Brett Esher, eds. *The Letters of Queen Victoria: A Selection from Her Majesty's Correspondence between the Years 1837 and 1861*. 3 vols. London: J. Murray, 1907.

Besant, Walter. *Autobiography of Sir Walter Besant*. London: Hutchinson, 1902.
– "The Late Walter Wren, Obituary." *The Times*, 9 August 1898, 8.
Besant, William Henry. "The Mathematical Tripos." In *The Student's Guide to the University of Cambridge*, edited by J.R. Seeley, 4–32. Cambridge: Deighton, 1880.
Best, Samuel. *Thoughts on the Proposals for the Improvement of the Civil Service; and for the Granting Diplomas through the Agency of the Institutions in Union with the Society of Arts*. London: Groombridge and Sons, 1853.
Blackie, Walter Graham. "Competitive Examinations for the Public Service, and the Preparatory Training in Our Scottish Universities." In *Transactions of the National Association for the Promotion of Social Science*, edited by George W. Hastings, 318–21. London: John W. Parker, 1860.
Bodichon, Barbara. "Middle-Class Schools for Girls." In *Transactions of the National Association for the Promotion of Social Science*, edited by George W. Hastings, 432–3. London: John W. Parker, 1860.
Bonython, Elizabeth, and Anthony Burton. *The Great Exhibitor: The Life and Work of Henry Cole*. London: V&A Publications, 2003.
Booth, James. *Examination the Province of the State: Or the Outlines of a Practical System for the Extension of National Education*. London: John W. Parker, 1847.
– *How to Learn and What to Learn*. London: Bell and Daldy, 1856.
Boulger, John. *The Master-Key to Public Offices, and Candidate's Complete Instructor*. London: Houlston and Wright, 1860.
Bourdieu, Pierre. *Outline of a Theory of Practice*. Cambridge: Cambridge University Press, 1977.
Bowker, Geoffrey. "Information Mythology: The World of/as Information." In *Information Acumen: The Understanding and Use of Knowledge in Modern Business*, edited by Lisa Bud-Frierman, 231–47. New York: Routledge, 1994.
"Brain-Power." *Chambers's Journal* 19 (1882): 649–52.
Brereton, Joseph Lloyd. *The Case for Examinations: An Account of their Place in Education with Some Proposals for Their Reform*. Cambridge: Cambridge University Press, 1944.
Briggs, William. *Intermediate Mathematics: A Guide to the Mathematics of the Intermediate Examinations in Arts and Science of the University of London*. London: W.B. Clive & Co., 1890.
Brock, William Hodson. "Geometry and the Universities: Euclid and His Modern Rivals, 1860–1901." *History of Education* 4 (1975): 21–35.
– *Queenwood College Revisited*. Aldershot: Ashgate Variorum, 1996.
– "School Science Examinations: Sacrifice or Stimulus?" In *Days of Judgement*, edited by Roy M. MacLeod, 169–88. Driffield: Studies in Education, 1982.
– "The Spectrum of Science Patronage." In *The Patronage of Science in the 19th Century*, edited by Gerard L'Estrange Turner, 173–206. Leiden: Noordhoff Publishing, 1976.
Browne, George Forrest. *The Recollections of a Bishop*. London: Smith, Elder, 1915.

Browning, Oscar. "Review: Report on the System of Education for the Middle and Upper Classes in France, Italy, Germany, and Switzerland." *Quarterly Review* 125 (1868): 473–89.

Bruce, Charles. *The Organization of Secondary and Superior Instruction ... In the Colony of Mauritius*. London: Edward Stanford, 1870.

[Bryant, Sophie]. "On the Maximum Utilisation of Examinations as an Aid to School Work." *Journal of Education* 5 (1883): 102–4.

[Burrows, Montagu]. "Report, Minutes of Evidence, & C., of the Schools Inquiry Commission." *Quarterly Review* 126 (1869): 448–79.

Busch, Lawrence. *Standards: Recipes for Reality*. Cambridge, MA: MIT Press, 2011.

Butterworth, Harry. "The Science and Art Department 1853–1900." PhD thesis, University of Sheffield, 1968.

– "The Science and Art Department Examinations: Origins and Achievements." In *Days of Judgement: Science, Examinations, and the Organization of Knowledge in Late Victorian England*, edited by Roy MacLeod, 27–43. Driffield: Studies in Education, 1982.

"The Calendar and General Directory of the Science and Art Department." *Nature* 35 (1887): 320–1.

Cambridge Local Examination Syndicate. *Examination of Students Who Are Not Members of the University*. Cambridge: University Press, 1859.

– *Local Examinations: Eighth Annual Report of the Syndicate, 16 Feb 1866*. Cambridge: University Press, 1866.

– *Report on an Examination of Girls, Held (by Permission of the Syndicate) in Connexion with the Local Examinations of the University of Cambridge in 1863*. London: Emily Faithfull, 1864.

Campbell, Carl. *Colony & Nation: A Short History of Education in Trinidad & Tobago 1834–1986*. Kingston, Jamaica: Ian Randle Publishers, 1992.

Caplan, Bryan. *The Case against Education: Why the Education System Is a Waste of Time and Money*. Princeton, NJ: Princeton University Press, 2018.

Caro, Robert Allan. *The Power Broker: Robert Moses and the Fall of New York*. New York: Knopf, 1974.

Carson, John. "Differentiating a Republican Citizenry: Talents, Human Science, and Enlightenment Theories of Governance." *Osiris* 17 (2002): 74–103.

– *The Measure of Merit: Talents, Intelligence, and Inequality in the French and American Republics, 1750–1940*. Princeton: Princeton University Press, 2007.

Caswell, Edward. *The Art of Pluck*. 9th ed. Oxford: J. Vincent, 1851.

[Chadwick, Edwin]. "A General, Medical, and Statistical History of the Present Condition of Public Charity in France." *London Review* 1, no. 2 (1829): 536–56.

– *A Lecture on the Economical, Social, Educational, and Political Importance of Open Competitive Examinations for Admission to the Public Service*. London: Knight and Co., 1857.

– "Open Competitive Examinations." *Journal of the Society of Arts* 6, no. 307 (8 October 1858): 671–3.

Chambers, Theodore Gervase. *Register of the Associates and Old Students of the Royal College of Chemistry, the Royal School of Mines, and the Royal College of Science*. London: Hazell, Watson and Viney, 1896.

Chester, Harry. "On the Society of Arts' Union of Institutes, and the Examinations Connected Therewith." *Journal of the Society of Arts* 7, no. 326 (18 February 1859): 192–6.

Cheth [pseud.]. "A Competitive Examination." *Uppingham School Magazine* 4 (September, October, November 1866): 69–74, 106–11, 212–20.

"China and the Chinese." *Westminster Review* 67 (1857): 526–57.

Civil Service Commissioners. *Third Report*. Parliamentary Paper C.185. London: Eyre and Spottiswoode, 1858.

– *Fifteenth Report*. Parliamentary Paper C.197. London: Eyre and Spottiswoode, 1870.

Civil Service Department. *Guide to the Examinations with Directions for Candidates*. London: P.S. King, 1856.

Clarendon Committee. *See* Select Committee of House of Lords on Public Schools Bill.

Clark, William. *Academic Charisma and the Origins of the Research University*. Chicago: University of Chicago Press, 2006.

Clarke, James Fernandez. *Autobiographical Recollections of the Medical Profession*. London: J. & A. Churchill, 1874.

Clarke, Richard Frederick. *The Influence of Pass Examinations: With a Scheme for Their Incorporation into the Honour Schools*. Oxford: James Parker, 1869.

Clough, Anne Jemima. "The Progress of Woman." *Universal Review* 2 (1888): 289–310.

Cole, Henry. *The Functions of the Science and Art Department*. London: Chapman and Hall, 1857.

Cole, Henry, and Alan Summerly Cole. *Fifty Years of Public Work of Sir Henry Cole*. 2 vols. London: George Bell and Sons, 1884.

"The Collapse of Wallahism." *Belgravia: A London Magazine* 6 (1875): 540–4.

College of Preceptors. *Fifty Years of Progress in Education: A Review of the Work of the College of Preceptors*. London: Printed for private circulation, 1897.

"College of Preceptors – August Distribution of Prizes and Certificates." *Educational Times*, August 1879, 215–19.

"College of Preceptors. Dean's Report." *Educational Times*, February 1862, 245.

"College of Preceptors – March Distribution of Prizes and Certificates." *Educational Times*, March 1879, 81–3.

Collins, Harry. *Gravity's Kiss: The Detection of Gravity Waves*. Cambridge, MA: MIT Press, 2017.

Collins, Randall. *The Credential Society: An Historical Sociology of Education and Stratification*. New York: Academic Press, 1979.

Commission on the State of Popular Education in England [Newcastle Commission]. *Report and Appendix to Minutes of Evidence*. Parliamentary Papers 2794-I. London: Eyre and Spottiswoode, 1861.

Committee of the Privy Council on Education. *Minute Establishing a Revised Code of Regulations*. Parliamentary Papers 2924. London: Eyre and Spottiswoode, 1862.

– *Minutes*. Printed annually.

– *Report*. Printed annually.

"A Competitive Examination Hall in China." *Once a Week* 7 (1871): 323–5.

Council of the University of London. *Second Statement, Explanatory of the Plan of Instruction*. London: John Murray, 1828.

Craik, Alex. *Mr Hopkins' Men: Cambridge Reform and British Mathematics in the 19th Century*. London: Springer, 2007.

Crockford, John. *Crockford's Scholastic Directory for 1861*. London: John Crockford, 1861.

Cumberlege, Geoffrey, ed. *The North London Collegiate School 1850–1950: A Hundred Years of Girls' Education*. Oxford: Oxford University Press, 1950.

Dale, William. *The Present State of the Medical Profession in Great Britain and Ireland*. London: Privately printed, 1860.

Dalgleish, Walter Scott. "University Certificate Examinations, or Local Examinations for Scotland." *Educational Times*, November 1863, 190–1.

– "University Certificate Examinations, or 'Local Examinations,' in Scotland." In *Transactions of the National Association for the Promotion of Social Science*, edited by George W. Hastings, 274–80. London: Longman, Green, Longman, Roberts, and Green, 1864.

Danvers, Frederick Charles et al. *Memorials of Old Haileybury College*. London: Archibald Constable, 1894.

Daston, Lorraine, and Peter Galison. *Objectivity*. New York: Zone, 2007.

Daunton, Martin. *Royal Mail*. London: Athlone Press, 1985.

David, Paul. "Clio and the Economics of Qwerty." *American Economic Review* 75, no. 2 (1985): 332–7.

Davies, Emily. *The Application of Funds to the Education of Girls*. London: Longman, Green, Longman, Roberts, and Green, 1865.

– *Special Systems of Education for Women*. London: Savill, Edwards, 1868.

de Carteret-Bisson, Frederick Shirley Dumaresq. *Our Schools and Colleges, for Boys*. 8th ed. Vol. 1. London: Simpkin, Marshall, 1884.

– *The Oxford and Cambridge Local Examination Record*. Vol. 2. London: Simpkin, Marshall, 1876.

[de Morgan, Augustus]. "Review of How to Learn and What to Learn." *The Athenaeum*, 13 December 1856.

Department of Science and Art. *Directory*. Printed annually.

– *Précis of the Minutes of the Science and Art Department from 16th Feb 1852 to 1st July 1863*. London: Eyre and Spottiswoode, 1869.

– *Report*. Printed annually. London: Eyre and Spottiswoode.

Delve, Janet. "The College of Preceptors and the Educational Times: Changes for British Mathematics Education in the Mid-Nineteenth Century." *Historia Mathematica* 30 (2003): 140–72.

de Salvo, Anna. *The Rise and Fall of the University Correspondence College.* Cambridge: National Extension College, 2001.

Desmond, Adrian. *Huxley: From Devil's Disciple to Evolution's High Priest.* London: Penguin, 1997.

– *The Politics of Evolution: Morphology, Medicine, and Reform in Radical London.* Chicago: University of Chicago Press, 1989.

– "Redefining the X Axis: 'Professionals,' 'Amateurs' and the Making of Mid-Victorian Biology: A Progress Report." *Journal of the History of Biology* 34 (2001): 3–50.

Desrosières, Alain. *The Politics of Large Numbers: A History of Statistical Reasoning.* Translated by Camille Naish. Cambridge, MA: Harvard University Press, 1998.

de Verteuil, Anthony. *The Holy Ghost Fathers of Trinidad.* Port of Spain: Litho Press, 1996.

Devonshire Commission. *See* Royal Commission on Scientific Instruction and the Advancement of Science (C.536)

Dickens, Charles. "Hard Times, Chapter 1." *Household Words: A Weekly Journal,* 1 April 1854, 141–71.

– *Sketches by Boz.* Leipzig: Tauchnitz, 1843. First published 1834.

"Discontinuance of the Educational Examinations of the Society of Arts." *Educational Times,* September 1879, 255.

"The Dispute between 'Crammers' and Public Schools." *Educational Times,* February 1880, 44, 61.

Dodds, William. *A Complete Guide to Matriculation at the University of London.* Manchester: John Heywood, 1871.

Donnelly, John Fretchville Dykes. *The Aid Now Granted by the State towards the Instruction of the Industrial Classes in Elementary Science: Its Nature and Results.* London: Department of Science and Art, 1861.

– *Lecture on the Promotion of Science Instruction by the Department of Science and Art.* London: Eyre and Spottiswoode, 1861.

Douglas, Heather. "The Irreducible Complexity of Objectivity." *Synthese* 138, no. 3 (2004): 453–73.

Duke, Christopher. "Robert Lowe – a Reappraisal." *British Journal of Educational Studies* 14, no. 1 (1965): 19–35.

Dyhouse, Carol. *No Distinction of Sex? Women in British Universities 1870–1939.* London: UCL Press, 1995.

[Eastlake, Elizabeth]. "Reports Issued by the Schools' Enquiry Commission on the Education of Girls." *Quarterly Review* 146 (1878): 40–69.

Eddy, Matthew. "The Interactive Notebook: How Students Learned to Keep Notes during the Scottish Enlightenment." *Book History* 19, no. 1 (2016): 86–131.

– "The Shape of Knowledge: Children and the Visual Culture of Literacy and Numeracy." *Science in Context* 26 (2013): 215–45.

Edgeworth, Francis Ysidro. "The Element of Chance in Competitive Examinations." *Journal of the Royal Statistical Society* 53 (1890): 460–75, 644–63.

– "The Uncertainty of Examinations." *Journal of Education* 12 (1 February 1890): 95–6.

"Education in the British Colonies." *Educational Times*, September 1878, 231–4.

"Educational and Literary Summary of the Month." *Educational Times*, May 1861, 43.

"Educational and Literary Summary of the Month." *Educational Times*, September 1863, 129–30.

"Educational Quacks." *London Review of Politics, Society, Literature, Art, and Science* 14 (1867): 449–50.

"Educational Summary of the Month." *Educational Times*, October 1879, 282–3.

"Educational Summary of the Month." *Educational Times*, June 1880, 159–60.

"Educational Summary of the Month." *Educational Times*, November 1880, 279.

Edwards, Paul. "Meteorology as Infrastructural Globalism." *Osiris* 21, no. 1 (2006): 229–50.

"Eighth Annual Conference." *Journal of the Society of Arts* 7, no. 346 (8 July 1859): 567–79.

Elman, Benjamin. *Civil Examinations and Meritocracy in Late Imperial China.* Cambridge, MA: Harvard University Press, 2013.

Elwick, James. "Economies of Scales: Evolutionary Naturalists and the Victorian Examination System." In *Victorian Scientific Naturalism: Community, Identity, Continuity*, edited by Gowan Dawson and Bernard Lightman, 131–56. Chicago: University of Chicago Press, 2014.

"The Encouragement of Scientific Research." *Quarterly Journal of Science* 6 (1876): 467–94.

Endersby, Jim. *Imperial Nature: Joseph Hooker and the Practices of Victorian Science.* Chicago: University of Chicago Press, 2008.

– "Odd Man Out: Was Joseph Hooker an Evolutionary Naturalist?" In *Victorian Scientific Naturalism: Community, Identity, Continuity*, edited by Gowan Dawson and Bernard Lightman, 2014. Chicago: University of Chicago Press, 2014.

Epictetus. *Discourses and Selected Writings.* Translated by Robert Dobbin. Penguin Classics, 2008.

Eschbach, Elizabeth Seymour. *The Higher Education of Women in England and America, 1865–1920.* New York: Garland, 1993.

Espeland, Wendy Nelson, and Michael Sauder. "Rankings and Reactivity: How Public Measures Recreate Social Worlds." *American Journal of Sociology* 113, no. 1 (2007): 1–40.

– *Engines of Anxiety: Academic Rankings, Reputation, and Accountability.* New York: Russell Sage, 2016.

Espeland, Wendy Nelson, and Mitchell L. Stevens. "Commensuration as a Social Process." *Annual Review of Sociology* 24 (1998): 313–43.

"Examination of Members of Classes in Institutions." *Journal of the Society of Arts* 4, no. 190 (11 July 1856): 587–90.

"Examinations." *Journal of the Society of Arts* 6, no. 286 (14 May 1858): 400–1.

An Examinee [pseud.]. "Notes in the Senate House." *Uppingham School Magazine* 5 (May 1867): 77–83.

"Experiences of a 'Little-Go' Coach." *Girton Review* 2 (July 1882): 6–8.

Eve, Henry Weston. "On Marking." In *Three Lectures on Subjects Connected with the Practice of Education*, edited by Henry Weston Eve, 1–28. Cambridge: University Press, 1883.

Ezrahi, Yaron. *The Descent of Icarus: Science and the Transformation of Contemporary Democracy*. Cambridge, MA: Harvard University Press, 1990.

Farley, John. *Gametes and Spores: Ideas about Sexual Reproduction, 1750–1914.* Baltimore: Johns Hopkins University Press, 1982.

Farnell, Charles Anson. *The Examination: Before, at, and After.* Manchester: J.B. Ledsham, 1885.

Faulkner, Robert, and Howard Becker. *Do You Know: The Jazz Repertoire in Action*. Chicago: University of Chicago Press, 2009.

Finer, Sheldon. *The Life and Times of Sir Edwin Chadwick*. London: Methuen, 1952.

A First Class Man [pseud.]. "Viva Voce Examinations." *The Times*, 17 August 1855, 10.

"First Ordinary Meeting." *Journal of the Society of Arts* 4, no. 209 (21 November 1856): 1–7.

Fitch, Joshua Girling. "Educational 'Results' and the Mode of Testing Them." In *Transactions of the National Association for the Promotion of Social Science*, edited by George W. Hastings, 258–67. London: John W. Parker and Sons, 1862.

– "Lectures on Practical Teaching." *Educational Times*, December 1878, 307–8.

– "Statistical Fallacies respecting Public Instruction." *Fortnightly Review* 14 (1873): 615–28.

[Fitch, Joshua Girling]. "The London University Calendar for 1857." *London Quarterly Review* 9 (1857): 1–22.

Foden, Frank. *The Examiner: James Booth and the Origin of Common Examinations.* Leeds: Leeds Studies in Adult and Continuing Education, 1989.

– *Philip Magnus: Victorian Educational Pioneer*. London: Vallentine, Mitchell, 1970.

– "Technical Examinations in England." *Paedagogica Historica: International Journal of the History of Education* 6, no. 1 (1966): 68–97.

Forgan, Sophie, and Graeme Gooday. "Constructing South Kensington: The Buildings and Politics of T.H. Huxley's Working Environments." *British Journal for the History of Science* 29, no. 4 (1996): 435–68.

Fortescue, Earl. *Public Schools for the Middle Classes*. London: Longman, Green, Longman, Roberts, and Green, 1864.

Foster, Peter le Neve. "Secretary's Report." *Journal of the Society of Arts* 6, no. 292 (25 June 1858): 492–4.

– "Letter to Secretary of Oldham Lyceum." *Journal of the Society of Arts* 6, no. 294 (9 July 1858): 537.

Foucault, Michel. *Discipline and Punish: The Birth of the Prison*. Translated by Alan Sheridan. London: A. Lane, 1977.

– *Power/Knowledge: Selected Interviews and Other Writings, 1972–1977*. Translated by Colin Gordon. New York: Pantheon Books, 1980.

– *Society Must Be Defended: Lectures at the Collège de France, 1975–76*. Translated by David Macey. New York: Picador, 1997.

Foucault, Michel, and Michel Senellart. *The Birth of Biopolitics: Lectures at the Collège de France, 1978–79*. Translated by Graham Burchell. Basingstoke: Palgrave Macmillan, 2008.

Fowler, Thomas. "On Examinations." *Fortnightly Review* 19, no. 111 (1876): 418–29.

A Fresh Surgeon [pseud.]. "New Examinations at the London College of Surgeons." *Lancet* 29, no. 748 (13 December 1837): 495.

Friedland, Martin L. *The University of Toronto: A History*. 2nd ed. Toronto: University of Toronto Press, 2013.

Frith, Simon. "Socialization and Rational Schooling: Elementary Education in Leeds before 1870." In *Popular Education and Socialization in the Nineteenth Century*, edited by Phillip McCann, 67–92. London: Methuen, 1977.

Froude, James Anthony. "Suggestions on the Best Means of Teaching English History." In *Oxford Essays, Contributed by Members of the University*, 47–79. London: John W. Parker and Son, 1855.

Fry, Herbert. *Our Schools and Colleges*. 1st ed. London: Robert Hardwicke, 1867.

– *Our Schools and Colleges*. 2nd ed. London: Robert Hardwicke, 1868.

Fyfe, Aileen, and Bernard V. Lightman. "Introduction." In *Science in the Marketplace: Nineteenth-Century Sites and Experiences*, edited by Aileen Fyfe and Bernard V. Lightman, 1–22. Chicago: University of Chicago Press, 2007.

Galton, Francis. *Hereditary Genius: An Enquiry into Its Laws and Consequences*. London: Macmillan, 1869.

Gardner, Alice. *A Short History of Newnham College, Cambridge*. Cambridge: Bowes & Bowes, 1921.

Gascoigne, John. "Mathematics and Meritocracy: The Emergence of the Cambridge Mathematical Tripos." *Social Studies of Science* 14 (1984): 547–84.

Geertz, Clifford. "Thick Description: Toward an Interpretive Theory of Culture." In *The Interpretation of Cultures: Selected Essays*, 3–32. New York: Perseus Books, 1973.

Gerson, Elihu. "The American System of Research: Evolutionary Biology, 1890–1950." PhD thesis, University of Chicago, 1998.

– "Reach, Bracket, and the Limits of Rationalized Coordination: Some Challenges for CSCW." In *Resources, Co-Evolution and Artifacts: Theory in CSCW*, edited by Mark Ackerman et al., 193–220. Springer, 2008.

Gigerenzer, Gerd. "From Tools to Theories: A Heuristic of Discovery in Cognitive Psychology." *Psychological Review* 98, no. 2 (1991): 254–67.

Gilchrist Educational Trust and University of London. *Conditions for Scholarships Instituted by the Gilchrist Educational Trust for the Benefit of Youth Residing in the Dominion of Canada*. London, 1879.

Ginzburg, Carlo. "Morelli, Freud and Sherlock Holmes – Clues and Scientific Method." *History Workshop Journal* 9 (1980): 6–36.

"Girton College." *Educational Times*, June 1880, 161–2.

"The Goffin Case." *Educational Times*, October 1879, 284.

Goffin, Robert Edward Hemblington. "Mr. Goffin's Certificate." *The Times*, 8 August 1879, 8.

Goldman, Lawrence. "Victorian Social Science: From Singular to Plural." In *The Organisation of Knowledge in Victorian Britain*, edited by Martin Daunton, 87–137. Oxford: Oxford University Press, 2005.

Golinski, Jan. *Making Natural Knowledge: Constructivism and the History of Science*. Cambridge: Cambridge University Press, 1998.

Gooday, Graeme. "Precision Measurement and the Genesis of Physics Teaching Laboratories in Victorian Britain." PhD thesis, University of Kent at Canterbury, 1989.

Goodwin, Harvey. *Memoir of Bishop Mackenzie*. Cambridge: Deighton, 1864.

– "The Sacrifice of Education." *Nineteenth Century* 25 (1889): 312–22.

Goody, Jack. *The Domestication of the Savage Mind*. Cambridge: Cambridge University Press, 1977.

Gordon, Shirley Courtnay. *A Century of West Indian Education*. London: Longmans, 1963.

Gould, Paula. "Femininity and Physical Science in Britain, 1870–1914." PhD thesis, University of Cambridge, 1998.

Government of Madras. "Resolution of the Government of Madras [1886]." In *Papers Relating to Technical Education in India 1886–1904*, 116–17. Calcutta, India: Office of the Superintendent of Government Printing, 1906.

Green, Simon. "Archbishop Frederick Temple on Meritocracy, Liberal Education and the Idea of a Clerisy." In *Public and Private Doctrine: Essays in British History Presented to Michael Cowling*, edited by Michael Bentley, 149–67. Cambridge: Cambridge University Press, 1993.

Grigg, Henry Bidewell. "Scheme for the Development of Scientific and Technical Education in Madras." In *Papers Relating to Technical Education in India 1886–1904*, 103–16. Calcutta, India: Office of the Superintendent of Government Printing, 1885.

Hacking, Ian. "Biopower and the Avalanche of Printed Numbers." *Humanities in Society* 5, no. 3–4 (1982): 279–95.

– *Historical Ontology*. Cambridge, MA: Harvard University Press, 2002.
– "Looping Effects of Human Kinds." In *Causal Cognition: A Multidisciplinary Debate*, edited by Dan Sperber, David Premack, and Ann James Premack, 351–94. Oxford: Oxford University Press, 1995.
Hansen, Hal. "Rethinking Certification Theory and the Educational Development of the United States and Germany." *Research in Social Stratification and Mobility* 29 (2011): 31–55.
Harris, Mary. *Common Threads: Women, Mathematics and Work*. Stoke on Trent: Trentham, 1997.
Hart, Jenifer. "The Genesis of the Northcote-Trevelyan Report." In *Studies in the Growth of Nineteenth-Century Government*, edited by Gillian Sutherland, 63–81. London: Routledge & Kegan Paul, 1972.
Harte, Negley. *The University of London, 1836–1986: An Illustrated History*. London: Athlone Press, 1986.
Hearl, Trevor. "Military Examinations and the Teaching of Science, 1857–1870." In *Days of Judgement: Science, Examinations, and the Organization of Knowledge in Late Victorian England*, edited by Roy M. MacLeod, 109–49. Driffield: Studies in Education, 1982.
Heller, Michael. *London Clerical Workers, 1880-1914: Development of the Labour Market*. London: Pickering & Chatto 2011.
Herbert, Auberon. "The Sacrifice of Education to Examinations." *Nineteenth Century* 24, 25 (November 1888, February 1889): 617–52, 284–322.
– ed. *The Sacrifice of Education to Examinations: Letters from "All Sorts and Conditions of Men."* London: Williams and Norgate, 1889.
– "State Education: A Help or Hindrance?" *Fortnightly Review* 28 (1880): 42–57.
Hill, Rowland. *Post Office Reform: Its Importance and Practicability*. London: Charles Knight, 1837.
"Hindoo Civil Servants." *All the Year Round* 2 (1869): 372–8.
Holman, Henry. *English National Education: A Sketch of the Rise of Public Elementary Schools in England*. London: Blackie and Sons, 1898.
Hooper, Henry. "College Life at Cambridge." *Westminster Review* 35 (1841): 456–80.
Hopkinson, David. "The Arnolds and Elementary Education." *The Arnoldian: A Review of Mid-Victorian Culture* 9, no. 1 (1982): 23–31.
Horn, Pamela. "The Corps of Royal Engineers and the Growth of British Technical Education." *Journal of Educational Administration and History* 14, no. 1 (1982): 18–26.
Hoskin, Keith. "The 'Awful Idea of Accountability': Inscribing People into the Measurement of Objects." In *Accountability: Power, Ethos and the Technologies of Managing*, edited by Rolland Munro and Jan Mouritsen, 265–82. London: International Thomson Business Press, 1996.
– "The Examination, Disciplinary Power and Rational Schooling." *History of Education* 8 (1979): 135–46.

Hoskin, Keith, and Richard Macve. "Accounting and the Examination: A Genealogy of Disciplinary Power." *Accounting, Organizations and Society* 11 (1986): 105–36.

Howarth, Janet. "Introduction." In *The Higher Education of Women*, edited by Janet Howarth, vii–xlvi. London: Hambledon Press, 1988.

Huber, V.A. *The English Universities: An Abridged Translation*. Translated by Francis W. Newman. 2 vols. London: William Pickering, 1843.

Hudson, Geoffrey Francis. *Europe & China; a Survey of Their Relations from the Earliest Times to 1800*. London: E. Arnold, 1931.

Hughes, Edward. "Sir Charles Trevelyan and Civil Service Reform, 1853–5." *English Historical Review* 64 (1949): 53–88.

Hutchins, Edwin. *Cognition in the Wild*. Boston: MIT Press, 1995.

Hutchinson, Alexander Hadden. *Guide to the Army-Competitive Examinations*. London: Edward Stanford, 1861.

Huxley, Thomas Henry. *Lessons in Elementary Physiology*. 2nd ed. London: Macmillan, 1868.

– "Science and Education." In *Collected Essays*. Vol. 3. London: Macmillan, 1895.

– "Technical Education [1877]." In *Science & Education: Collected Essays*, 404–26. London: MacMillan, 1899.

[Huxley, Thomas Henry]. "National Education in Science and Art." *The Times*, 31 January 1887.

Huxley, Thomas Henry, and Leonard Huxley. *Life and Letters of Thomas Henry Huxley*. 2nd ed. 3 vols. London: Macmillan, 1908.

Indian Civil Service. *The Selection and Training of Candidates for the Indian Civil Service*. London: Eyre and Spottiswoode, 1876.

Jackson, Ian. "Zuleika and the Moslem." *Book Collector* 41, no. 1 (1992): 74–7.

Jacobs, Andrea. "Girls & Examinations 1860–2002." PhD thesis, University of Southampton (Winchester University), 2003.

– "'The Girls Have Done Very Decidedly Better Than the Boys': Girls and Examinations 1860–1902." *Journal of Educational Administration and History* 33 (2001): 120–36.

Jarrell, Richard. *Educating the Neglected Majority: The Struggle for Agricultural and Technical Education in Nineteenth-Century Ontario and Quebec*. Montreal: McGill-Queen's University Press, 2016.

– "Visionary or Bureaucrat? T.H. Huxley, the Science and Art Department and Science Teaching for the Working Class." *Annals of Science* 55 (1998): 219–40.

Jex-Blake, Sophia. "Women as Practitioners of Midwifery." *Lancet* 92:2445 (1870): 63–4.

– "The Medical Education of Women." In *Transactions of the National Association for the Promotion of Social Science*, edited by C.W. Ryalls, 385–93. London: Longmans, Green, 1874.

– *Medical Women: A Thesis and a History*. 2nd ed. Edinburgh: Oliphant, Anderson, & Ferrier, 1886.

Johns, Adrian. *The Nature of the Book: Print and Knowledge in the Making*. Chicago: University of Chicago Press, 1998.

Jones, Claire. "Femininity and Mathematics at Cambridge circa 1900." In *Women, Education, and Agency, 1600–2000*, edited by Jean Spence, Sarah Jane Aiston, and Maureen Meikle, 147–67. London: Routledge, 2010.

Jones, Gareth Stedman. "The Determinist Fix: Some Obstacles to the Further Development of the Linguistic Approach to History in the 1990s." *History Workshop Journal* 42 (1996): 19–35.

Kafka, Ben. *The Demon of Writing: Powers and Failures of Paperwork*. New York: Zone Books, 2012.

Kay-Shuttleworth, Sir James. *Four Periods of Public Education*. London: Longman, Green, Longman, and Roberts, 1862.

Keenan, Patrick. *Papers on the State of Education in Trinidad*. HMSO, 1870.

Keetley, Charles Bell. *The Student's Guide to the Medical Profession*. London: Macmillan, 1878.

"King Cram in India." *All the Year Round* 8 (1872): 232–5.

Kuhn, Thomas. *The Essential Tension: Selected Studies in Scientific Tradition and Change*. Chicago: University of Chicago Press, 1977.

Kula, Witold. *Measures and Men*. Translated by Richard Szreter. Princeton, NJ: Princeton University Press, 1986.

A Lady Principal [pseud.]. "Life and Work in a Girls' School." *Golden Hours: An Illustrated Monthly Magazine for Family and General Reading* (1872): 247–52.

Lakoff, George, and Mark Johnson. *Metaphors We Live By*. 2nd ed. Chicago: University of Chicago Press, 2003.

Lampland, Martha, and Susan Leigh Star. "Reckoning with Standards." In *Standards and Their Stories: How Quantifying, Classifying, and Formalizing Practices Shape Everyday Life*, edited by Martha Lampland and Susan Leigh Star, 3–24. Ithaca, NY: Cornell University Press, 2009.

[Lankester, Edwin Ray]. "Instruction to Science Teachers at South Kensington." *American Naturalist* 5, no. 11 (1871): 685–93.

Latham, Henry. *On the Action of Examinations Considered as a Means of Selection*. Cambridge: Deighton Bell, 1877.

Latour, Bruno. *Pandora's Hope: Essays on the Reality of Science Studies*. Cambridge, MA: Harvard University Press, 1999.

– *Science in Action: How to Follow Scientists and Engineers through Society*. Milton Keynes: Open University Press, 1987.

– "Visualization and Cognition: Thinking with Eyes and Hands." In *Knowledge and Society: Studies in the Sociology of Culture Past and Present*, edited by Henrika Kuklick and Elizabeth Long, 1–39. Greenwich, CT: JAI Press, 1986.

Layton, David. *Science for the People*. London: George Allen & Unwin, 1973.

Leys, Ruth. *From Sympathy to Reflex: Marshall Hall and His Opponents.* New York: Garland, 1990.

"Life at a 'Crammer's.'" *Chambers's Journal* 792 (1879): 139–41.

Light, Jennifer. "When Computers Were Women." *Technology and Culture* 40, no. 3 (1999): 455–83.

Lightman, Bernard V. *Victorian Popularizers of Science: Designing Nature for New Audiences.* Chicago: University of Chicago Press, 2007.

"Liverpool Petition for Medical Reform." *Lancet* 2, no. 833 (17 August 1839): 754–5.

Locke, John. "Of the Conduct of the Understanding." In *Works*, 331–402. London: C. and J. Rivington, 1824.

Lowe, Robert. *Middle Class Education: Endowment or Free Trade.* London: Robert John Bush, 1868.

– "Shall We Create a New University?" *Fortnightly Review* 21 (1877): 160–71.

Lowell, Abbott Lawrence. *Colonial Civil Service: The Selection and Training of Colonial Officials in England, Holland, and France.* London: Macmillan, 1900.

Macaulay, Thomas, Lord Ashburton, Henry Melvill, Benjamin Jowett, and John George Shaw Lefevre. "Report on the Indian Civil Service." In *Sessional Papers of the House of Lords, 1854–5 – Miscellaneous Subjects.*

MacDonagh, Oliver. "The Nineteenth-Century Revolution in Government: A Reappraisal." *Historical Journal* 1 (1958): 52–67.

MacKenzie, Donald A. *An Engine, Not a Camera: How Financial Models Shape Markets.* Cambridge, MA: MIT Press, 2006.

MacLeod, Roy. *Days of Judgement: Science, Examinations, and the Organization of Knowledge in Late Victorian England.* Driffield: Studies in Education, 1982.

– "Introduction: Science and Examinations in Victorian England." In *Days of Judgement: Science, Examinations, and the Organization of Knowledge in Late Victorian England*, edited by Roy M. MacLeod, 2–24. Driffield: Studies in Education, 1982.

Mann, Horace [London]. "On the Recent Report of the Select Committee of the House of Commons on the Civil Service Appointments." In *Transactions of the National Association for the Promotion of Social Science*, edited by George W. Hastings, 302–11. London: John W. Parker, 1860.

[Mann, Horace]. [Boston, MA]. "Boston Grammar and Writing Schools." *Common School Journal* 7 (1845).

Martin, Arthur Patchett. *Life and Letters of the Right Honourable Robert Lowe.* London: Longmans, Green, 1893.

Martin, William. "Competitive Examinations in China." *North American Review* 111 (1870): 62–78.

Mason, Donald. "Peelite Opinion and the Genesis of Payment by Results: The True Story of the Newcastle Commission." *History of Education* 17, no. 4 (1988): 269–81.

McWilliams-Tullberg, Rita. *Women at Cambridge: A Men's University – Though of a Mixed Type*. London: Gollancz, 1975.

Melville, David. "The Prize System in Education." In *Transactions of the National Association for the Promotion of Social Science*, edited by George W. Hastings, 249–57. London: John W. Parker, 1858.

"Medical Studies!" *London University Magazine* 3 (1858): 136–9.

"Middle Class Examinations." *English Journal of Education* 11 (new series) (1857): 225–9.

Mill, Anna. "Some Notes on Mill's Early Friendship with Henry Cole." *Mill News Letter* 4, no. 2 (1969): 2–13.

Mill, John Stuart. "Endowments." *Fortnightly Review* 5 (1869): 377–90.

Mills, Charles Wright. *The Power Elite*. Oxford: Oxford University Press, 1956.

Mirowski, Philip. "Looking for Those Natural Numbers: Dimensionless Constants and the Idea of Natural Measurement." *Science in Context* 5, no. 1 (1992): 165–88.

Morris, Norman. "An Historian's View of Examinations." In *Examinations and English Education*, edited by Stephen Wiseman, 1–44. Manchester: Manchester University Press, 1961.

Moses, Robert. *The Civil Service of Great Britain*. New York: Columbia University Press, 1914.

"Mr. Gladstone on Education and Competitive Examinations." *Educational Times*, May 1862, 40–1.

"Mr. Golightly; or, Memoirs of a Cambridge Freshman, Chapter 19." *Once a Week* 7 (1 April 1871): 338–43.

Mundella, Anthony John. "Introducing the Proposals for a New Code." *Practical Teacher* 1 (1881): 332–8.

Newcastle Commission. *See* Commission on the State of Popular Education in England.

Norris, John Pilkington. *The Education of the People, Our Weak Points and Our Strength, Occasional Essays*. London: Simpkin, Marshall, 1869.

– "On the Proposed Examination of Girls of the Professional and Middle Classes." In *Transactions of the National Association for the Promotion of Social Science*, edited by George W. Hastings, 404–12. London: Longman, Green, Longman, Roberts and Green, 1864.

Northcote, Stafford Henry. "Competitive Examination for the Civil Service." In *Transactions of the National Association for the Promotion of Social Science*, edited by George W. Hastings, 279–86. London: John W. Parker, 1859.

"Obituary – Mr. Walter Wren." *The Times*, 8 August 1898.

Occasional Papers on University Matters. Cambridge: Macmillan, 1858.

"Organization of the Civil Service." *Journal of the Society of Arts* 2, no. 67 (3 March 1854): 273–7.

Ova [pseud.]. "The Examinee: A Reminiscence." *Ulula: The Manchester Grammar School Magazine* 11 (December 1883): 34–7.

"Oxford Associate-in-Arts Examination." *Chambers's Journal* 246 (1858): 187–9.

An Oxford BA [Hugh Chisholm]. "University Degrees for Women." *Fortnightly Review* 57 (1895): 895–903.

"Oxford Local Examination." *Educational Times*, January 1864, 221.

"Oxford Local Examinations." *Educational Times*, September 1863, 123–4.

Oxford University Commission. *Inquiry into the State, Discipline, Studies, and Revenues of the University and Colleges of Oxford*. London: W. Clowes and Sons, Stamford, 1852. First published 1482.

An Oxonian [pseud.]. "Almae Matres. No. III. – University Education." *Titan: A Monthly Magazine* 26 (1858): 565–85.

Paley, William. *Works*. London: Henry G. Bohn, 1849.

Papers and Correspondence between Science and Art Dept. and Committee of Class No. 3150, Relating to Examination, May 1878. London, 1879.

Papers on the Re-organisation of the Civil Service. Parliamentary Papers 1870. London: Eyre and Spottiswoode, 1855.

Papillon, Thomas Leslie. "The Late Walter Wren, Obituary." *The Times*, 10 August 1898.

Pascoe, Charles Eyre. *Schools for Girls and Colleges for Women: A Handbook of Female Education Chiefly Designed for the Use of Persons of the Upper Middle Class*. London: Hardwicke and Bogue, 1879.

Pedersen, Joyce Senders. *The Reform of Girls' Secondary and Higher Education in Victorian England: A Study of Elites and Educational Change*. New York: Garland, 1987.

Petley, Brian William. *The Fundamental Physical Constants and the Frontier of Measurement*. Bristol: Adam Hilger, 1985.

Philo-Veritas [pseud.]. "Examination of a Student of the London University at the College of Surgeons." *Lancet* 13, no. 336 (30 January 1830): 645–6.

"Piece-Work at South Kensington." *Punch* 44 (14 March 1863): 104.

Playfair, Lyon. *Industrial Instruction on the Continent*. London: Eyre and Spottiswoode, 1852.

– *On Teaching Universities and Examining Bodies*. 3rd ed. Dublin: Hodges, 1873.

– *Subjects of Social Welfare*. London: Cassell & Company, 1889.

Pollock, Frederick. *Oxford Lectures, and Other Discourses*. London: MacMillan, 1890.

Poovey, Mary. *Genres of the Credit Economy: Mediating Value in Eighteenth- and Nineteenth-Century Britain*. Chicago: University of Chicago Press, 2008.

Porter, Theodore. "How Science Became Technical." *Isis* 100, no. 2 (2009): 292–309.

– "Quantification and the Accounting Ideal in Science." *Social Studies of Science* 22 (1992): 633–52.

– "Thin Description: Surface and Depth in Science and Science Studies." *Osiris* 27, no. 1 (2012): 209–26.

– *Trust in Numbers: The Pursuit of Objectivity in Science and Public Life.* Princeton, NJ: Princeton University Press, 1995.

Post Office. *Account of the Celebration of the Jubilee of Uniform Inland Penny Postage.* London: Richard Clay and Sons, 1891.

Preston-Thomas, Herbert. *The Work and Play of a Government Inspector.* Edinburgh: W. Blackwood, 1909.

"Programme of Examinations for 1860." *Journal of the Society of Arts* 7, no. 351 (12 August 1859): 627–31.

"Prospectus of University Correspondence College Classes for the Examinations of the University of London." In *Xenophon's Oeconomicus, a Translation*, 2–16. Cambridge: University Correspondence College, 1890.

Quick, Robert Hebert. "Examination Papers." *Educational Times*, April 1878, 97–101.

Reathlous, Henry. "Matriculation Hints." *Practical Teacher* 4 (1885): 93–4, 139–40, 173, 212, 376–7, 567–8.

"Recent Examination Questions." *Practical Teacher* 4 (1884): 47.

Reeks, Margaret. *Register of the Associates and Old Students of the Royal School of Mines.* London: Royal School of Mines Old Students' Association, 1920.

Reese, William. *Testing Wars in the Public Schools: A Forgotten History.* Cambridge, MA: Harvard University Press, 2013.

"Registry of Candidates Who Have Gained Certificates." *Journal of the Society of Arts* 4, no. 200 (19 September 1856): 705.

Reidy, Michael. *Tides of History: Ocean Science and Her Majesty's Navy.* Chicago: University of Chicago Press, 2008.

Renwick, Chris. "The Practice of Spencerian Science: Patrick Geddes's Biosocial Program, 1876–1889." *Isis* 100 (2009): 36–57.

"Reviews and Notices." *Educational Times*, May 1861, 40.

"Rewards and Honours for Proficiency in Science." *Popular Science Review* I (1862): 126–7.

Richmond, Marsha. "'A Lab of One's Own': The Balfour Biological Laboratory for Women at Cambridge University, 1884–1914." *Isis* 88 (1997): 422–55.

– "The 1909 Darwin Celebration: Reexamining Evolution in the Light of Mendel, Mutation, and Meiosis." *Isis* 97 (2006): 447–84.

Rigg, James. "Minute of the Committee of Privy Council on Education, Establishing a Revised Code of Regulations." *London Review* 17 (1862): 583–624.

Roach, John. *Public Examinations in England, 1850–1900.* Cambridge: Cambridge University Press, 1971.

Roberts, John Anthony George. *A Concise History of China.* Cambridge, MA: Harvard University Press, 1999.

Rodger, Nicholas Andrew Martin. *The Command of the Ocean: A Naval History of Britain, 1649–1815.* New York: W.W. Norton, 2005.

Rolleston, George John, and William Turner. *Scientific Papers and Addresses*. Oxford: Clarendon, 1884.

Romanes, George. "Mental Differences between Men and Women." *Nineteenth Century* 123 (1887): 654–72.

Roscoe, Henry Enfield. *The Life and Experiences of Sir Henry Enfield Roscoe*. London: MacMillan, 1906.

Rose, Nikolas. "Calculable Minds and Manageable Individuals." *History of the Human Sciences* 1, no. 2 (1988): 179–200.

Rothblatt, Sheldon. "The Student Sub-Culture and the Examination System in Early 19th Century Oxbridge." In *The University in Society*, edited by Lawrence Stone, 247–304. Princeton, NJ: Princeton University Press, 1974.

– "Supply and Demand: The 'Two Histories' of English Education." *History of Education Quarterly* 28, no. 4 (1988): 627–44.

Rowe, William. *China's Last Empire: The Great Qing*. Cambridge, MA: Belknap Press, 2009.

Royal Commission on Scientific Instruction and the Advancement of Science [Devonshire Commission]. *First and Second Report, Minutes of Evidence, Appendices*. C.536. London: Eyre and Spottiswoode, 1872.

Royal Commission on Technical Instruction [Samuelson Commission]. *Second Report*. Parliamentary Papers Ch. 3981. London: Eyre and Spottiswoode, 1884.

Royal Commission to Inquire into Education in Schools in England and Wales [Taunton Commission]. *Minutes of Evidence* (pt. 1, pt. 2). Parliamentary Papers 3966-III, 3966-IV. London: Eyre and Spottiswoode, 1867–8.

– *Report* (pt. 1, pt. 2). Parliamentary Papers 3966. London: Eyre and Spottiswoode, 1867–8.

Rupke, Nicolaas. *Richard Owen: Victorian Naturalist*. New Haven: Yale University Press, 1994.

Russell, Bertrand. *Portraits from Memory, and Other Essays*. London: George Allen & Unwin, 1956.

Sadler, Michael. "The Scholarship System in England to 1890 and Some of Its Developments." In *Essays on Examinations*, 1–78. London: Macmillan, 1936.

Sadler, Michael, and Jack Sislian. *Representative Sadleriana: Sir Michael Sadler (1861–1943), on English, French, German, and American School and Society*. New York: Nova Science Publishers, 2004.

Samuelson Commission. *See* Royal Commission on Technical Instruction

Samuelson Committee. *See* Select Committee on Scientific Instruction

Samuelson, James. "The Progress of Science Schools and Classes; with Hints for Their Formation." *Popular Science Review* 1 (1862): 223–8.

Sayce, Archibald Henry. "Results of the Examination-System at the Universities." *Fortnightly Review* 17 (1875): 835–46.

Schaffer, Simon. "Metrology, Metrication, and Victorian Values." In *Victorian Science in Context*, edited by Bernard Lightman, 438–74. Chicago: University of Chicago Press, 1997.

– "Modernity and Metrology." In *Science and Power: The Historical Foundations of Research Policies in Europe*, edited by Luca Guzzetti, 71–92. Luxembourg: Office for Official Publications of the European Communities, 2000.

Scharlieb, Mary. "The Progress of Woman in Science and Medicine." *Universal Review* 2 (1888): 310–16.

"Schedules. Standards of Examination in Elementary and Class Subjects." *Practical Teacher* 1 (1881): 331.

"Scholastic Registration." *Educational Times*, January 1862, 219.

Schwartz, Leonard. "Professions, Elites and Universities in England, 1870–1970." *Historical Journal* 47 (2004): 941–62.

"Science Training in Elementary Schools." *Journal of the Society of Arts* 27, no. 1361 (1878): 63–70.

Scott, James. *Seeing Like a State: How Certain Schemes to Improve the Human Condition Have Failed*. New Haven, CT: Yale University Press, 1998.

Secord, James. "Knowledge in Transit." *Isis* 95 (2004): 654–72.

– *Victorian Sensation: The Extraordinary Publication, Reception, and Secret Authorship of Vestiges of the Natural History of Creation*. Chicago: University of Chicago Press, 2000.

Sedbergh School. *Register, 1546 to 1895*. Leeds: Richard Jackson, 1895.

Seivad, Hadji [Gerald Stanley Davies]. *The Moslem in Cambridge*. Cambridge: Metcalfe and Sons, 1890 [1870].

Select Committee. *Report to Inquire into Circumstances Relating to Suspension of Mr. Goffin's Certificate by Science and Art Department*. Parliamentary Papers 334. London: Eyre and Spottiswoode, 1879.

Select Committee of House of Lords on Public Schools Bill [Clarendon Committee]. *Report*. Parliamentary Papers 481. London: HMSO, 1865.

Select Committee on Medical Education [Warburton Committee]. *Report, Minutes of Evidence*. Parliamentary Papers 602. 1834.

Select Committee on Medical Registration. *Report*. Parliamentary Papers 620. London: House of Commons, 1847.

Select Committee on Postage. *First Report*. Parliamentary Papers 278. 1838.

Select Committee on Scientific Instruction [Samuelson Committee]. *Report*. Parliamentary Papers 432, 432-I. London: HMSO, 1868.

"Seventh Annual Conference." *Journal of the Society of Arts* 6, no. 293 (2 July 1858): 501–10.

Sherborne, Michael. *H.G. Wells: Another Kind of Life*. London: Peter Owen, 2010.

Shteir, Ann. *Cultivating Women, Cultivating Science: Flora's Daughters and Botany in England, 1760–1860*. Baltimore: Johns Hopkins University Press, 1996.

Siklos, Stephen. *Philippa Fawcett and the Mathematical Tripos*. Cambridge: Newnham College, 1990.

Simon, Josep. *Communicating Physics: The Production, Circulation and Appropriation of Ganot's Textbooks in France and England, 1851–1887*. London: Pickering & Chatto, 2011.

– "Secondary Matters: Textbooks and the Making of Physics in Nineteenth-Century France and England." *History of Science* (2012): 339–74.

Smith, David, ed. *The Correspondence of H.G. Wells*. Vol. 1, *1880–1903*. London: Pickering & Chatto, 1998.

Smith, Frank. *The Life and Work of Sir James Kay-Shuttleworth*. Bath: Cedric Chivers, 1974.

Sneyd-Kynnersley, Edmund McKenzie. *H.M.I.: Some Passages in the Life of One of H.M. Inspectors of Schools*. London: MacMillan, 1913.

"Statistics of Education in England and Wales." *The Times*, 28 March 1861, 8.

Stephen, Leslie. *Sketches from Cambridge*. London: Macmillan, 1865.

Stinchcombe, Arthur L. *When Formality Works: Authority and Abstraction in Law and Organizations*. Chicago: University of Chicago Press, 2001.

Stockwell, Anthony John. "Examinations and Empire: The Cambridge Certificate in the Colonies, 1857–1957." In *Making Imperial Mentalities: Socialization and British Imperialism*, edited by J.A. Mangan, 203–19. Manchester: Manchester University Press, 1990.

Stoddart, Anna. *Life and Letters of Hannah E. Pipe*. Edinburgh: William Blackwood, 1908.

Strathern, Marilyn. "Accountability ... and Ethnography." In *Audit Cultures: Anthropological Studies in Accountability, Ethics, and the Academy*, edited by Marilyn Strathern, 279–304. London: Routledge, 2000.

Stray, Christopher. *Classics Transformed: Schools, Universities, and Society in England, 1830–1960*. Oxford: Clarendon Press, 1998.

– "Educational Publishing." In *The History of Oxford University Press*, edited by Ian Gadd, Simon Eliot, and W. Roger Louis, 2:472–510. Oxford University Press, 2013.

– *Grinders and Grammars: A Mid-Victorian Controversy*. Reading: Textbook Colloquium, 1995.

– *Oxford Classics: Teaching and Learning, 1800–2000*. London: Duckworth, 2007.

– "The Shift from Oral to Written Examination: Cambridge and Oxford 1700–1900." *Assessment in Education: Principles, Policy & Practice* 8 (2001): 33–50.

– "Zulu and Zuleika." *Book Collector* 42, no. 3 (1993): 429–31.

Stray, Christopher, and Gillian Sutherland. "Mass Markets: Education." In *The Cambridge History of the Book in Britain*, edited by David McKitterick, 359–81. Cambridge: Cambridge University Press, 2009.

Sturges, Octavius, and Mary Sturges. *In the Company's Service, a Reminiscence*. London: W.H. Allen, 1883.

Sutherland, Gillian. "Anne Jemima Clough and Blanche Athena Clough: Creating Educational Institutions for Women." In *Practical Visionaries: Women, Education and Social Progress 1790–1930*, edited by Mary Hilton and Pam Hirsch, 101–14. Harlow: Longman, 2000.

– "Education." In *Cambridge Social History of Britain*, edited by Francis Michael Longstreth Thompson, 119–69. Cambridge: Cambridge University Press, 1990.

– *Elementary Education in the Nineteenth Century*. London: The Historical Association, 1971.

– "Emily Davies, the Sidgwicks and the Education of Women in Cambridge." In *Cambridge Minds*, edited by Richard Mason, 34–47. Cambridge: Cambridge University Press, 1994.

– "Examinations and the Construction of Professional Identity: A Case Study of England 1800–1950." *Assessment in Education: Principles, Policy & Practice* 8 (2001): 51–64.

– *Faith, Duty and the Power of Mind: The Cloughs and Their Circle 1820–1960*. Cambridge: Cambridge University Press, 2006.

– *Policy-Making in Elementary Education, 1870–1895*. London: Oxford University Press, 1973.

Suum Cuique [pseud]. "The Indian Civil Service." *The Times*, 6 August 1862, 6.

Sylvester, David William. *Robert Lowe and Education*. Cambridge: Cambridge University Press, 1974.

Taunton Commission. *See* Royal Commission to Inquire into Education in Schools in England and Wales

Taylor, R.V. "Cambridge Local Examinations." *Educational Times*, May 1865, 84–5.

Temple, Frederick. *Middle Class Examinations: Report of the Results of the West of England Examination*. Vol. 3. Exeter: W.H. Roberts, 1857.

Teng, Ssu-yu. "Chinese Influence on the Western Examination System: I. Introduction." *Harvard Journal of Asiatic Studies* 7 (1943): 267–312.

Thorburn, John. *Female Education from a Physiological Point of View*. Manchester: Owens College, 1884.

Thorold, Algar Labouchere. *The Life of Henry Labouchere*. New York: G.P. Putnam's Sons, 1913.

Tilly, Charles. *Popular Contention in Great Britain, 1758–1834*. Cambridge, MA: Harvard University Press, 1995.

– *Social Movements, 1768–2004*. Boulder: Paradigm Publishers, 2004.

Todhunter, Isaac. *The Conflict of Studies and Other Essays on Subjects Connected with Education*. London: Macmillan, 1873.

– *The Elements of Euclid for the Use of Schools and Colleges; Comprising the First Six Books and Portions of the Eleventh and Twelfth Books*. London: Macmillan, 1862.

Topham, Jonathan. "Scientific Publishing and the Reading of Science in Nineteenth-Century Britain: A Historiographical Survey and Guide to Sources." *Studies in History and Philosophy of Science* 31 (December 2000): 559–612.

Trevelyan, Charles. *The Irish Crisis*. London: Longman, Brown, Green & Longmans, 1848.
Trevelyan, George Otto. *The Competition Wallah*. London: Macmillan, 1866.
– *The Life and Letters of Lord Macaulay*. 2 vols. London: Longmans, Green, 1876.
Trollope, Anthony. *The Three Clerks*. 3 vols. London: Richard Bentley, 1858.
Trollope, Anthony, and David Skilton. *An Autobiography*. London: Penguin Books, 1996.
Tropp, Asher. *The School Teachers: The Growth of the Teaching Profession in England and Wales from 1800 to the Present Day*. Westport, CT: Greenwood Press, 1977.
Tyndall, John. *Testimonials of John Tyndall, Ph.D., Candidate for the Professorship of Natural Philosophy in the University of Toronto*. London: Richard Taylor, 1851.
United Westminster Schools. *Annual*. London: W.H. & A. Nutting, 1880.
University of Cambridge. *Examination of Students Who Are Not Members of the University*. Cambridge: University Press, 1859.
University of London. *Minutes of Committees. December 2nd 1847 to December 31st 1850*. Privately printed, 1847–50.
– *Minutes of Committees, 1853–1866*. Privately printed, 1853–66. University of London Archives ST 3/2/6.
– *Minutes of the Senate March 4th 1837 to December 21st 1843*. Vol. 1. London: Taylor and Francis, 1837–43.
– *Minutes of the Senate June 22nd 1843 to December 31st 1849*. Vol. 2. London: Taylor and Francis, 1843–9.
– *Minutes of the Senate*. London: Richard Taylor, 1853–8.
– *Minutes of the Senate 1855–1858*. Vol. 4. London: Taylor and Francis, 1855–8.
– *Minutes of the Senate 1859–1862*. Vol. 5. London: Taylor and Francis, 1859–62.
– *Minutes of the Senate 1879–1882*. Vol. 10. London: Taylor and Francis, 1879–82.
– *Papers Relating to the Origin of the University*. 1857.
– *The University of London: The Historical Record*. London: University of London Press, 1912.
Uzzell, Peter. "The Science and Art Department and the Teaching of Chemistry." *The Vocational Aspect of Education* 29, no. 74 (1977): 127–32.
Wallace, Alfred Russel. *My Life: A Record of Events and Opinions*. 2 vols. London: Chapman & Hall, 1905.
United Kingdom. 19 *Parliamentary Debates* H.C. 3rd ser. (1833), cols. 469–550.
– 158 *Parliamentary Debates* H.C. 3rd ser. (1860), cols. 880–970.
– 264 *Parliamentary Debates* H.C. 3rd ser. (1881), cols. 1193–1345.
– 128 *Parliamentary Debates* L. 3rd ser. (1853), cols. 1–57.
Warburton Committee. *See* Select Committee on Medical Education
Warwick, Andrew. *Masters of Theory: Cambridge and the Rise of Mathematical Physics*. Chicago: University of Chicago Press, 2003.

– "A Mathematical World on Paper: Written Examinations in Early 19th Century Cambridge." *Studies in History and Philosophy of Modern Physics* 29 (1998): 295–319.
– "'A Very Hard Nut to Crack' or Making Sense of Maxwell's Treatise on Electricity and Magnetism in Mid-Victorian Cambridge." In *Scientific Authorship: Credit and Intellectual Property in Science*, edited by Mario Biagioli and Peter Galison, 133–65. London: Routledge, 2003.
Weber, Max. *Economy and Society: An Outline of Interpretive Sociology*. Edited by Roth Guenther and Wittich Claus. 2 vols. Berkeley and Los Angeles: University of California Press, 1978.
"Well-Known Teachers at Work: William Briggs." *Practical Teacher* 13 (1893): 384–88.
Wells, Herbert George. *Experiment in Autobiography*. Toronto: Macmillan, 1934.
– *Love and Mr. Lewisham*. London: Collins, 1966. First published 1899.
– *The New Machiavelli*. London: John Lane, 1911.
– *Text-Book of Biology*. University Correspondence College Tutorial Series. London: W.B. Clive & Co., 1893.
"What Women Are Educated For." *Once a Week* 5 (1861): 175–9.
Whewell, William. *On the Principles of English University Education*. London: John W. Parker, 1837.
White, Paul. *Thomas Huxley: Making the 'Man of Science'*. Cambridge: Cambridge University Press, 2003.
Wiese, Luguy Adolf. *German Letters on English Education: Written during an Educational Tour in 1876*. Translated by Leonard Schmitz. London: William Collins, Sons, and Co., 1877.
Wilks, Samuel. "The Admission of Women to the University of London." *British Medical Journal* (23 May 1874): 695.
Williams, Frederick. *Our Iron Roads: Their History, Construction and Social Influences*. London: Ingram, Cooke, 1852.
Willis, Richard. "Market Forces and State Intervention in Educational Enterprise: The Case of School Examinations from 1850 to 1917." *Journal of Educational Administration and History* 27 (1995): 97–109.
Wilson, James Maurice. *James M. Wilson: An Autobiography 1836–1931*. London: Sidgwick & Jackson, 1932.
– "On Teaching Natural Science in Schools." In *Essays on a Liberal Education*, edited by F.W. Farrar, 241–92. London: Macmillan, 1867.
[Wilson, Robert]. "Aesculapia Victrix." *Fortnightly Review* 39 (1886): 18–33.
The Wire. "Know Your Place." Season 4, episode 9. Directed by Alex Zakrzewski. Written by David Simon. Home Box Office, 12 November 2006.
Wise, Norton. "Introduction." In *The Values of Precision*, edited by Norton Wise, 3–16. Princeton: Princeton University Press, 1995.

Wolfram, Henry. *The Private Tutor's "Raison D'être."* London: Edward Stanford, 1885.

Wood, Henry Trueman. *A History of the Royal Society of Arts*. London: John Murray, 1913.

"A Word for 'Cram.'" *London Review*, 18 August 1866, 179–80.

"The Working of the Revised Code." *Museum and English Journal of Education* 3 (January 1868): 366–9.

Workman, W.P. "How to Write an Examination Paper." *Practical Teacher* 1 (1881): 10–12.

Wren, Walter. "Education and Examination." *Universal Review* 2 (1888): 391–402.

Young, Michael Dunlop. *The Rise of the Meritocracy, 1870–2033*. London: Thames & Hudson, 1958.

Yuan, Chen. "My Thoughts on the Funeral of Mr. Sun Yat Sen." *Xiandai pinglun* 1, no. 18 (11 April 1925): 4–5.

Zeng, Kangmin. *Dragon Gate: Competitive Examinations and Their Consequences*. London: Cassell, 1999.

Index

www.ingramcontent.com/pod-product-compliance
Lightning Source LLC
LaVergne TN
LVHW040151080826
844660LV00014B/924/J

* 9 7 8 1 4 8 7 5 0 8 9 3 7 *